Coasts

Peter Stiff

Advanced
TopicMaster

Series editor
Michael Raw

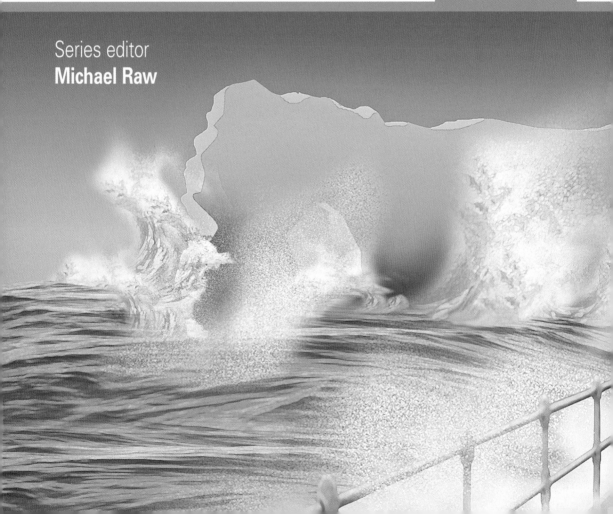

Acknowledgements

I would like to thank Michael Morrish and Michael Raw for the photographs they have supplied. In addition, Michael Raw's advice and encouragement as series editor and longstanding geographical 'companion' have been invaluable.

Some of the photographs were taken by my father, and to both him and my mother I owe a considerable debt of gratitude for all those traditional seaside holidays. They would have been fascinated to see where it has all led.

Philip Allan Updates, an imprint of Hodder Education, an Hachette UK company, Market Place, Deddington, Oxfordshire OX15 0SE

Orders

Bookpoint Ltd, 130 Milton Park, Abingdon, Oxfordshire, OX14 4SB

tel: 01235 827720

fax: 01235 400454

e-mail: uk.orders@bookpoint.co.uk

Lines are open 9.00 a.m.–5.00 p.m., Monday to Saturday, with a 24-hour message answering service. You can also order through the Philip Allan Updates website: www.philipallan.co.uk

ISBN 978-1-84489-615-8

First printed 2007
Impression number 5 4 3
Year 2012 2011 2010

Printed in Spain

Hachette UK's policy is to use papers that are natural, renewable and recyclable products and made from wood grown in sustainable forests. The logging and manufacturing processes are expected to conform to the environmental regulations of the country of origin.

P001387

Contents

Introduction

This book provides an up-to-date review of coasts and coastal management for students of AS/A-level geography. It is primarily aimed at sixth-form students, but it may also be useful for first-year undergraduates.

Why study the coastal zone? There are many reasons. First, it contains fascinating and diverse landforms and landscapes — from dramatic towering sea cliffs to quiet creeks meandering through salt marshes — which require explanation. Second, it has huge economic and ecological importance. Third, coasts and coastal management are key topics in AS and A-level specifications. And finally, because we live on a group of relatively small islands, we are forced to engage with the coast, which is never far away.

The coming together of land, sea and atmosphere makes for an ever changing environment. You only need to spend a short time at the coast to appreciate its dynamic nature — for example, the hour-by-hour progress of tides or the changes in wave conditions with the passage of a storm. There is, therefore, intellectual challenge and reward when studying coastal systems.

But quite apart from any academic justification, coasts also have great practical significance for millions of people. They are hazardous environments: cliff and dune erosion threaten coastal settlements, and the destruction of sea defences puts others at risk from flooding. Meanwhile, climate change and rising sea levels are likely to make the coast even more hazardous in future.

Sustainable management of the coastal zone requires an understanding of coastal processes and systems. This book begins with a study of the key inputs of energy and sediment into the coastal zone. It then looks at the principal types of coastal environment: those dominated by rocky landforms and those collectively known as beaches. The role of coastal ecosystems is investigated, as is the influential factor of sea level change. Once an understanding of these processes is acquired, human activity and management within the coastal zone are considered. As we will see, many of the failures of coastal management in the past stem from inadequate understanding of the coastal system. Recent approaches to management recognise the need to work with coastal systems rather than against them. As with river management, this new approach puts the emphasis on 'soft' rather than 'hard' structures. In some cases, it even means letting the natural processes take their course and abandoning (controversially) attempts to hold back the sea.

As a student, you can use this book as a learning tool in several ways. Most obviously, the book provides knowledge and understanding of how coasts operate and how people interact with them. The text should be read to consolidate your understanding simultaneously with coverage of each topic in class. You will, of course, need to refer to specific areas of the text to complete essays and other assignments.

Don't ignore the case studies. They are particularly important to support general discussion and explanation. They should be used to illustrate not only your assignments, but also your written work in final examinations. Information is backed up with photos, tables and diagrams. These should be scrutinised just as carefully as the text.

An integral feature of this book is its many activities, through which it becomes interactive. Some activities aim to test your knowledge and understanding; some develop essential skills such as statistical analysis and hypothesis testing; and others encourage you to investigate topics further, through research on the internet or in your library. For those who really engage with coasts, there are many ideas that could form the basis for fieldwork investigations and coursework.

Peter Stiff

1 Coasts as energy systems

Coasts occur where sea, land and atmosphere meet. This unique combination is a dynamic environment where change occurs frequently and at regular intervals. It produces some spectacular landscapes and landforms, and offers a variety of opportunities for human activity (Figure 1.1).

Figure 1.1 **Waves breaking on the seafront at Broadstairs**

Peter Stiff

Defining the coast is not always straightforward, although where tall cliffs plunge directly into the sea the boundary is clear. Other locations, such as extensive, low-lying estuaries, seem at times to be terrestrial, but change over the course of a few hours into a marine environment. The most useful context for a study of coasts is to think in terms of a **coastal zone**. This zone extends between the inland limit of coastal influence — for example, where sea spray can be blown — and the seaward limit of the land's influence, such as how far river water extends as an identifiable flow. However, not to be excluded are the marine processes that extend beyond such a zone, such as the circulation of sediment in an offshore cell.

Coasts as open systems

The coastal zone is a dynamic system. It is also an open system, with inputs and outputs of energy and materials (Figure 1.2).

Figure 1.2 **Energy inputs into the coastal system**

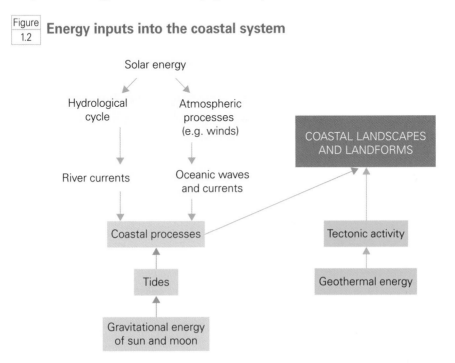

Within the same section of coast, stores of materials and energy can be found and a wide variety of processes operate. The coastal system is driven by:

- **solar energy**, which powers the hydrological cycle and winds
- **gravitational energy**, which is responsible for:
  - the attraction of sea water by the moon and sun (the tides)
  - the movement of material down slopes
  - underwater landslides that generate tsunamis
- **geothermal energy**, which is responsible for tectonic activity

Some of these energy inputs operate over long time scales (e.g. the convergence of plates along the west coast of South America), whereas winds can be generated and die away in a matter of hours.

Under the influence of energy, matter moves through the coastal system. Sediment, transported by waves and currents, enters a stretch of coast, sometimes passing through it in a relatively short period of time. Sediment can also remain within a coastal section as a sand dune, beach or offshore bar for years or even decades.

The end result of these inputs, stores and processes, is a variety of landforms and landscapes that, through the effects of erosion, transport and deposition, change continually.

Energy inputs: waves, tides and currents

A key input of energy to the coastal zone comes via moving water as waves, tides and currents. These have the potential to erode, transport and deposit materials.

Waves

Waves represent a transfer of energy and are capable of carrying out much work. They are described by:

- **height** — the vertical distance between wave crest and wave trough (Figure 1.3)
- **length** — the horizontal distance between consecutive crests (Figure 1.3)
- **wave period** — the time taken for consecutive crests to pass a fixed point
- **wave steepness** — the ratio of wave height to wavelength
- **wave velocity** — the ratio of wavelength to wave period

Figure 1.3 | **Wave characteristics**

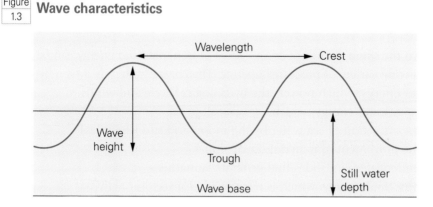

Wave generation

Most of the waves affecting coastal zones are **wind generated**. They are caused by the frictional drag between wind and water. The amount of energy transferred between the wind and the water depends on:

- wind velocity
- wind duration
- fetch (the distance over which the wind blows)

Wave height is commonly taken as an indication of wave energy. Two frequently used formulae for a measure of wave height are:

$$H = 0.031U^2$$
$$H = 0.36\sqrt{F}$$

where H = wave height (m), U = wind velocity (ms^{-1}) and F is the fetch (km). 0.031 and 0.36 are empirically derived constants.

The greatest energy transfer occurs when strong winds blow in the same direction over a long distance and for a long period of time; this produces the highest waves. On a global scale, it is possible to identify coasts exposed to different levels of wave energy on the basis of prevailing wind speeds, fetch and the aspect of the coast (Figure 1.4).

Figure 1.4 **Global distribution of coasts exposed to high wave energy**

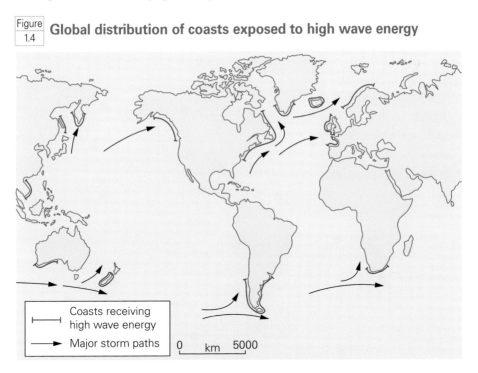

Coasts receiving high wave energy

Major storm paths

0 km 5000

Activity 1

Use a map at an appropriate scale to measure the maximum possible fetch for waves arriving at the British coast from the following directions: NE; NW; E; W; SE; SW and S.

(a) Using the formula

$$H = 0.36\sqrt{F}$$

calculate the maximum possible wave height at these locations, as determined by fetch.

Activity 1 (continued)

(b) Describe and explain the pattern of high-, medium- and low-wave-energy environments, as influenced by fetch.

(c) Comment on the influence of prevailing winds in relation to anticipated wave energy at these locations.

Activity 2

Figure 1.5 **Location of selected moored buoys, lightships and oil/gas platforms around the British Isles**

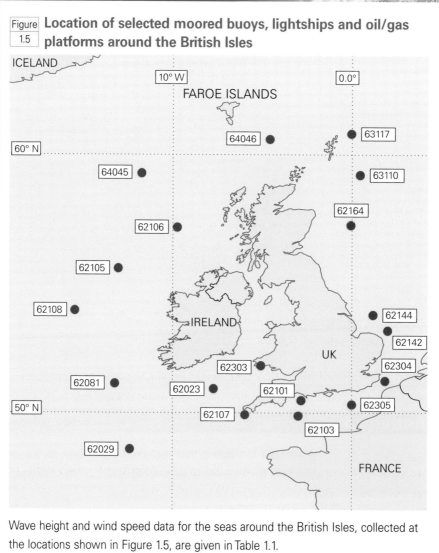

Wave height and wind speed data for the seas around the British Isles, collected at the locations shown in Figure 1.5, are given in Table 1.1.

Activity 2 (continued)

Table 1.1 Wave height and wind speed data for the seas around the British Isles

Data collection locations	Wind speed (m s⁻¹)	Wave height (m)
62023	15.6	4.0
62029	22.7	6.2
62081	14.4	5.7
62101	15.5	3.4
62103	15.4	3.3
62105	14.6	6.7
62106	14.7	6.7
62107	18.0	5.0
62108	14.9	6.1
62142	10.1	1.4
62144	10.2	1.1
62164	11.8	1.5
62303	13.4	4.2
62304	13.2	2.1
62305	16.5	2.4
63110	11.7	1.6
63117	12.8	4.1
64045	14.2	5.9
64046	13.2	5.7

(a) Use the data in Table 1.1 to draw a scattergraph that shows the relationship between wind speed and wave height.

(b) Analyse the statistical significance of the relationship between wind speed and wave height using a correlation coefficient.

(c) Comment on the result and its significance for energy input to the coastal zone.

(d) Describe and explain one factor, other than wind speed, that is likely to influence wave height.

Two other categories of wave can interact with the land.

Storm surges are generated by the combination of extremely high wind speed and very low atmospheric pressure. The latter allows water to rise well above normal levels, an effect that is accentuated when combined with a high tide. Surges occur regularly along coasts affected by tropical storms, most commonly in the Bay of Bengal. They can also occur around the coastline of the North Sea. The effects can extend a long way inland along low-lying shores.

Seismic sea waves (tsunamis) are generated when a large mass of water is displaced. Earthquakes and large-scale underwater mass movements can trigger tsunamis. In deep water, tsunamis are hardly noticeable, having a height that is usually less than 1m and a long wavelength, up to 200km. They travel at velocities of up to 600 km h^{-1}. Their impact is severe when they approach the coastal zone. There they can reach a height of many metres and so transfer their energy some way inland and with tremendous force.

For example, on 26 December 2004, movement along a subduction zone off the northern tip of Sumatra in the eastern Indian Ocean was registered as an earthquake of magnitude 9.3 on the Richter scale. The seabed in the vicinity was forced upwards by some 20m. This displaced vast quantities of water and set in motion high-energy waves that raced across the Indian Ocean. When the waves reached the shore closest to the epicentre in Aceh province, Indonesia, they were 20m high. It was estimated that 1000 tonnes of water crashed down on each metre of shore. More than 200 000 people lost their lives in the few hours it took for the tsunami to spread around the region.

Waves in deep water

There is little horizontal water movement in waves in the open ocean as waves are pulses of energy that travel through the water. Individual water particles move in a circular motion in waves in deep water, at an equal velocity in all parts of the orbit (Figure 1.6). These waves are known as **oscillatory waves** or **sea waves**. The diameter of the circle travelled by a water particle decreases with increasing depth of water until a depth is reached at which the water is unaffected by the pulses of energy. This depth is known as the **wave base**.

| Figure 1.6 | **Circular motion in deep-water waves** |

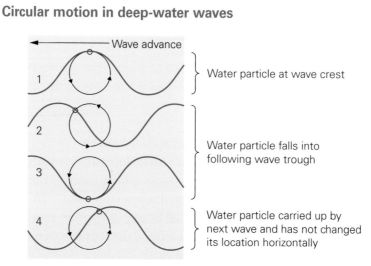

Although not a precise definition, wave base is generally acknowledged to be one-half to one-quarter of the wavelength.

Waves in shallow water

Waves bring vast amounts of energy into the coastal system and can, therefore, carry out considerable geomorphologic work. As they come close to the land, waves are modified as a result of decreasing water depth. The circular orbits of deep-water waves become more elliptical because of friction between the seabed and the water particles, with the forward element of the elliptical orbit being faster than the backward one. Wavelength and wave velocity both decrease. As energy has to be conserved, it is transferred into an increase in wave height and consequently wave steepness. A situation is reached eventually in which water is piled to a height that over-steepens the leading part of the wave, so that the orbits of the particles are broken and the water rushes forward as the wave 'breaks'. Once the wave has lost its shape, water moves up the shore as **swash** and returns to the sea under the force of gravity as **backwash** (Figure 1.7).

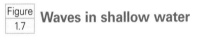

| Figure 1.7 | **Waves in shallow water** |

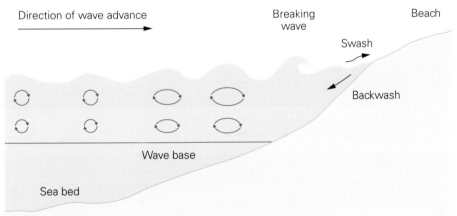

Types of breaking wave

Breaking waves are not all the same. The type of breaking wave depends on:
- wave steepness
- water depth
- the gradient of the shore

Breaking waves exist along a continuum, with various names given to different types of breaker (Table 1.2).

Table 1.2 Types of breaker

Breaker type	Description
Spilling (Figure 1.8)	Steep; low-angled shore gradient; break at some distance from the shore; foam forms at wave crest and becomes a line of surf as wave approaches shore
Plunging (Figure 1.9)	Steep; steep-angled shore gradient; steep-fronted; tend to curl over and plunge down onto shore
Surging (Figure 1.10)	Gentle; steep-angled shore gradient; tend not to break completely; top of wave breaks close to shore; water slides up and down the shore

Figure 1.8 Spilling breakers, Dawlish Warren, Exe estuary, Devon

Peter Stiff

Figure 1.9 Plunging breakers, Beer, east Devon

Peter Stiff

Figure
1.10 **Surging breakers, Mojacar, Costa de Almería, southern Spain**

Michael Raw

The significance of breaker type is the amount of energy that each brings into the coastal zone and how this impacts on the shore. Wave energy is dissipated by:

- breaking waves
- friction with the seabed
- erosion and transport of sediments
- erosion of solid rock

The terms **high-** and **low-energy breakers** are used to indicate how much work, in terms of erosion, transport and deposition, a breaker might carry out.

Activity 3

For each of the three breaker categories listed in Table 1.2, describe and explain possible impacts, in terms of the balance between erosion, transport and deposition, on a length of:

- sandy beach
- shingle beach

Wave refraction and diffraction

Wave refraction and diffraction are important because they affect the distribution of wave energy along the coast.

Wave refraction

Whenever waves approach the shore at an oblique angle, the part of the wave in shallower water closer to the shore decelerates due to friction with the seabed. The remainder of the wave, in water deeper than the wave base, moves forward at a constant speed. As a result, the wave bends (refracts) so that its orientation is more parallel to the coastline. An indication of this effect on energy distribution can be seen when lines drawn at right angles to wave crest (**wave rays**) are traced from offshore to the shore (Figure 1.11).

<table>
<tr><td>Figure
1.11</td><td>**Wave refraction and energy patterns**</td></tr>
</table>

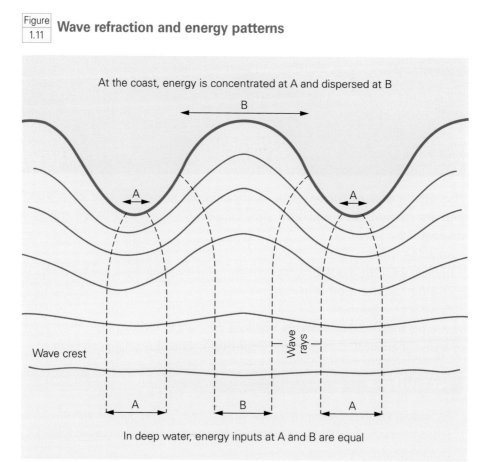

At the coast, energy is concentrated at A and dispersed at B

In deep water, energy inputs at A and B are equal

Along relatively straight stretches of coast, refraction is minimal, so energy distribution tends to be even along the shore. Waves approaching stretches of coastline that consist of headlands and bays are more subject to refraction (Figure 1.12). As a result, energy is focused on the headlands, where wave rays converge. Meanwhile, energy is more dispersed in the bays, where wave rays diverge.

Figure
1.12

Wave refraction in Runswick Bay, North Yorkshire

Michael Raw

Wave energy is also affected by the shape of the seabed in the coastal zone. The presence of canyons or shallow areas influences the pattern of wave energy arriving at the shore.

Activity 4

(a) Trace a stretch of coastline shown on an Ordnance Survey map (scale 1:50000 or 1:25000). Choose a coastline that faces between south to west.

(b) Plot a series of wave crests approaching from the southwest. As each successive wave crest is drawn closer to the shore, use the principles of refraction to distort the line.

(c) Mark the pattern of wave rays.

(d) Describe the pattern of high- and low-energy-receiving parts of the coastline. Suggest how this might affect the processes occurring along your chosen stretch of coast.

Wave diffraction

Wave diffraction occurs when a wave meets an obstacle, such as an island or an offshore breakwater around which it passes. Although the lee of the obstacle is protected from wave action, once the wave crest is past the obstacle, water flows into the sheltered area and wave action still occurs. This has implications for coastal management schemes seeking to protect coastlines from wave action.

Tides

Tides result from the gravitational attraction on water by the moon and the sun, the former having about twice the effect of the latter. Two important aspects of tides are:

- tidal frequency
- tidal range (the vertical distance between low and high tides)

Tidal frequency

Most coastlines experience **semi-diurnal tides** — two high and two low tides approximately every 24 hours. Some locations, for example California, experience a more varied pattern, where the two high (or low) tides can be quite different. Antarctica is unusual. It receives diurnal tides — one high tide and one low tide every 24 hours. As the respective motions of the Earth, moon and sun go through regular cycles, the gravitational forces change and, therefore, so do the tides. Twice a month the sun and the moon are aligned so that their gravitational forces are combined. This results in above-average tides called **spring tides**. Twice a month the sun and moon are at right angles with respect to the Earth. Their gravitational forces do not act in the same direction, so lower than average tides result. These are called **neap tides** (Figure 1.13).

| Figure 1.13 | **The formation of tides** |

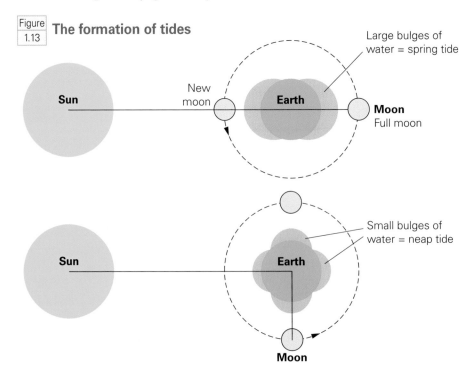

Tidal range

Tidal range is at its highest at the time of spring tides; neap tides give rise to the lowest range.

Tidal range varies considerably. Enclosed seas such as the Mediterranean experience hardly any difference in sea level between high and low tides. On the other hand, the physical shape of the coast, including the underwater topography, can amplify tidal oscillation, resulting in a tidal range of over 10m. The Bay of Fundy in northeast Canada has a range of 16m, the greatest in the world.

The significance of tidal range to a coastline is its effect on the extent of the inter-tidal zone and on the time interval between tides. These influence weathering processes and biological activity. With a high range, more of the coastal zone is exposed to the effects of wetting and drying on the rocks, and the waves operate over a greater vertical range, bringing more energy into the coastal zone. Coasts with a high range tend to possess greater biodiversity in the inter-tidal zone because more ecological niches occur. For example, a large tidal range is often associated with tidal flats and salt marshes.

Amphidromic points

A tide is a wave of energy governed largely by the moon, so it might be expected that the crest of this wave lies directly beneath the moon as it orbits the Earth. However, this is not the case because of such factors as:
- variations in ocean depth
- the seabed topography
- the shapes of the landmasses
- the ocean being broken up into deep basins (e.g. Atlantic and Pacific) separated by shallower shelves and continents

The overall effect is that the tide is broken up into smaller systems rather than one worldwide wave. As the gravitational pull of the moon passes across a basin, so does the tide. It is rather like holding a shallow dish full of water and gently tipping it from side to side. The water swirls around with a point in the centre (the **amphidromic point**) at which the water level hardly changes. The greatest changes in water depth take place furthest away from this point.

Because of the earth's rotation, tides circulate around an amphidromic point. Co-tidal lines join points at which high tide occurs at the same time. Co-tidal lines radiate from an amphidromic point and are usually drawn at one hour intervals starting at a particular point along a coast. High and low tides therefore occur at different times along a coastline as the water swirls around the amphidromic point.

On the global scale, there are six major amphidromic points where water levels change very little and around which tides move. For example, there are amphidromic points in the middle of the North Atlantic and halfway between South Africa and Antarctica. Smaller basins such as the North Sea also possess amphidromic points (Figure 1.14).

Figure 1.14 **Amphidromic points and tidal patterns in the North Sea**

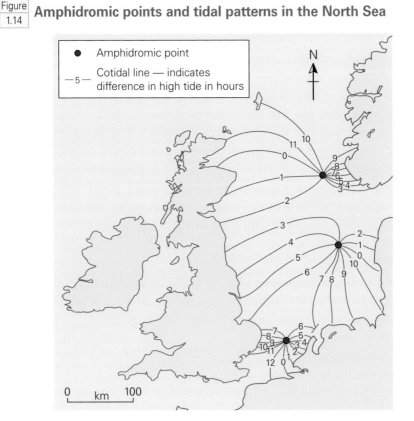

Currents

There are clearly identifiable water flows in the coastal zone:

- tidal currents
- shore-normal currents
- longshore currents
- rip currents
- river currents

The energy represented by tidal currents is significant in eroding, transporting and depositing material. In estuaries, the rising tide (known as the **flood tide**) can pick up (entrain) sediment and carry it inland. Once high tide is reached, the current reverses and the **ebb tide** takes over, carrying material in the opposite direction. Current velocities are relatively low at the start and end of each cycle but are at their maximum in the middle of the rising or falling tide. Different-sized particles are, therefore, entrained and deposited at different times and in different locations.

Where waves approach the shore with their crests parallel to the shape of the coastline, shore-normal currents exist. Water is carried up the beach, but there has to be a return flow. So, at fairly evenly spaced locations along the beach, rip currents flow back through the advancing waves at speeds of up to about $1\,\mathrm{m\,s^{-1}}$ (Figure 1.15).

Figure 1.15 **Rip currents**

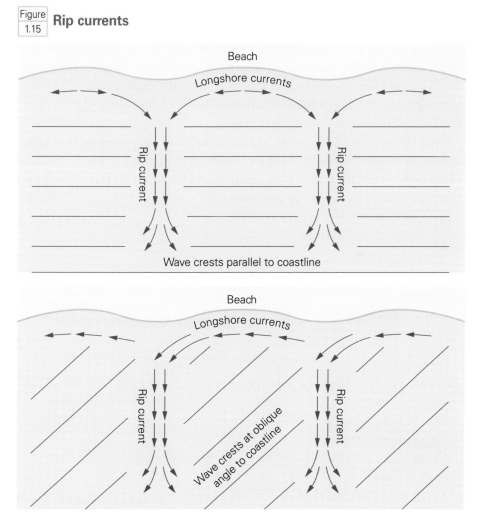

Where waves approach the shore at an angle (rarely greater than 10°) the predominant water movement is along the shore (longshore). Water eventually flows back out to sea as rip currents (Figure 1.15).

In estuaries, rivers transport both fluvially-derived sediment and fresh water into the coastal zone. These flows can be energetic and can affect coastal processes, such as sediment transport along the coast and the pattern of sediment deposition within estuaries.

Equilibrium and coasts

Given that the energy inputs to the coast are so dynamic, change in the processes operating at the coast and in the landforms making up the coast, is inevitable. Such changes occur across a range of time scales, from the effect of a single wave breaking on a beach to the impact of tectonic adjustments in geological time. Landforms develop a morphology (shape) that dissipates available energy. Too much energy leads to erosion and transport; too little, to deposition. As energy inputs change, landform shape is adjusted. Some changes take place in just a few hours; others take thousands of years.

Equilibrium is achieved when the amount of energy entering the coastal system is equal to the energy dissipated, without any change in the morphology of the landforms. This **steady-state** equilibrium (Figure 1.16(a)) continues until there is a change in the energy environment.

Figure 1.16	**Types of equilibrium**

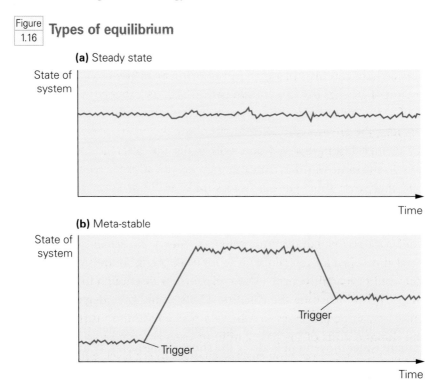

(a) Steady state

State of system

Time

(b) Meta-stable

State of system

Trigger

Trigger

Time

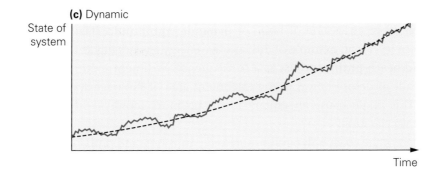

(c) Dynamic

State of system

Time

Where and when sea cliffs receive more or less the same atmospheric and marine energy (for example climate and waves), the profile of the cliff tends to stay the same from year to year. A beach receiving similar amounts of wave energy from one year to the next undergoes seasonal adjustments, but its average annual gradient stays the same.

Sometimes a dramatic event occurs within the coastal zone that brings about substantial change to the coast. Sand and shingle can adjust rapidly to changing energy inputs. A high-energy storm that generates increases in wave height and periodicity can remove most of a beach in the space of a few hours. The result is a new shape — a wide, flat beach. Subsequently, the coastal zone adjusts to the new situation and a **meta-stable** equilibrium is established (Figure 1.16(b)). Wave energy is absorbed without any further net transfer of sediment.

Human activity can also bring about adjustment in the coastal zone. The construction of groynes over a 6-month period triggers a change in the rate and volume of sediment moving along a coast. In turn, this influences the beaches and cliffs near the groynes and also further along the coast.

Energy change lasting a few hours will make no difference to solid-rock coastlines. In the same way that river channels are not adjusted when rivers flow over solid rock, over short periods, waves make no impression on solid-rock coastlines. Even if significant erosion occurs during a single storm, it will take thousands of years to achieve anything approaching an equilibrium form.

On hard-rock coasts, equilibrium could, in theory, be achieved if environmental stability lasted for long enough — for example, thousands of years. Cliff recession would eventually form a shore platform wide enough to dissipate all wave energy before reaching the cliff line. Cliffs would then degrade through weathering and slope processes to achieve a new equilibrium form. However, given the climatic shifts of the past 2 million years as ice sheets advanced and retreated, and the accompanying eustatic changes in sea level, equilibrium on such a time scale is unlikely.

Within the context of long-term change, changes can be more gradual. The coast could be in a state of basic or **dynamic** equilibrium (Figure 1.16(c)). Rising sea level is an example of this: as wave energy reaches higher up the shore, cliff and beach profiles adjust as a consequence. Sediment input can increase as a result of deforestation in the catchment area of the river draining to the coast. This might cause an estuary to silt up more rapidly, with salt marsh growth accelerating.

So, part of the coast is in equilibrium and part is not, and probably never will be. The system has freedom to adjust to energy changes on short time scales by shifting sediments (similar to a river flowing across its floodplain within a channel made up from alluvium) but little, if any, to bring about equilibrium on rocky coasts. As a result, in stormy conditions, the coastal system has surplus energy. Much of this surplus is expended battering cliffs. The evidence for this is erosional features formed by cliff recession — for example, caves, arches and stacks.

We must not forget that wave energy also has an uneven spatial distribution on indented coastlines. Refraction concentrates wave energy on headlands but diffuses it in bays. Energy environments can change in just a few metres, adding further complexity to the coastal system. Thus, variations in energy inputs are spatial as well as temporal.

2 Rocky coastal environments

Where a high level of energy and relatively resistant geology are found together in the coastal zone, a rocky coastline develops. The most spectacular landforms are cliffs, but a range of other landforms may be present. Rocky coastal environments are found throughout the world. However, some of the more prominent rocky stretches are where an active plate boundary is also a coastline. The effect of significant tectonic activity, for example earthquakes, can be to cause sustained vertical movement. This creates a narrow continental shelf that allows high levels of wave energy to reach the shore. Much of North and South America have extensive lengths of rocky coastline because these conditions are satisfied.

Processes

When there is a high level of wave energy, marine processes can be significant, as can bioerosion and a variety of sub-aerial processes.

Marine erosion

Wave action is the most important erosive force acting in the coastal zone. Its effect varies with wave energy and with the coastal geology. When steep cliffs plunge straight down into relatively deep water, waves tend not to break before they reach the cliff. In this case, hardly any erosion occurs because there is little forward movement of water and the waves reflect back from the cliff. It is more common to find coastlines where waves break and some of the energy is used to erode the coast (Figure 2.1). Marine erosion consists of the following processes:
- hydraulic action
- abrasion
- attrition

Hydraulic action

Hydraulic action includes processes that result from the movement of water without the involvement of rock particles. The impact of waves causes variations in pressure due to the weight of the water and its movement. Rock that is alternately covered by water and then exposed to the air becomes weakened by the fluctuating forces acting on it. Added to these forces are **pneumatic stresses** associated with moving air. Water and air can be trapped and compressed between forward moving water and a cliff face. This is particularly the case when the cliff is made of well-jointed rock. As the wave retreats, the pressure is suddenly released, which weakens the cliff face. In addition, the impact of a mass of water can dislodge fractured and loose rock.

Figure 2.1 **Waves breaking on recent basalt flow, Hawaii**

Peter Stiff

The term **wave quarrying** is used when loose blocks of rock are eroded. If the waves are extremely large, **cavitation** occurs. Air bubbles, at great pressure within the wave, collapse. This generates shock waves that erode rock surfaces with a similar effect to hammer blows. Small-scale features such as flutes and grooves can also result from cavitation.

Abrasion

Breaking waves pick up and carry sediment such as sand, gravel and pebbles. As the moving water drags the sediment over rock and as sediment is thrown at a rock face, a scouring action takes place, known as **abrasion** (also called corrasion). The effectiveness of abrasion depends on wave energy and on the availability of sediment. Larger sediment (boulders and pebbles) are moved only when the water has plenty of energy, for example in an intense storm. Smaller sediment (sand and gravel) is moved more frequently and by lower-energy waves. For example, along a short stretch of the North Yorkshire coast near Whitby, cliff erosion increases by a factor of 15 where sediment is available.

Attrition

Individual sediment particles collide with each other as they are moved around by water. When this happens, fragments are broken off, which reduces the size of the particles. Only smaller sediment is carried as suspended load in the water, with the larger particles being rolled up and down the beach as successive waves break, run up the beach and then run back down towards the sea. This movement produces smoothed and rounded sediment — a process known as **attrition**. Sediment is also broken and rounded when it is involved in abrasion.

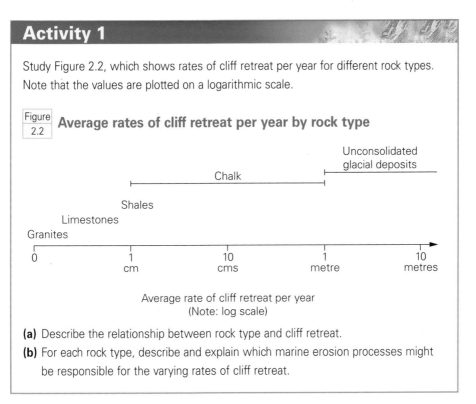

Activity 1

Study Figure 2.2, which shows rates of cliff retreat per year for different rock types. Note that the values are plotted on a logarithmic scale.

Figure 2.2 **Average rates of cliff retreat per year by rock type**

Average rate of cliff retreat per year
(Note: log scale)

(a) Describe the relationship between rock type and cliff retreat.

(b) For each rock type, describe and explain which marine erosion processes might be responsible for the varying rates of cliff retreat.

Bioerosion

Bioerosion is caused by the activity of biological organisms.

- Gastropods and echinoids rasp a rock surface as they graze over it, which gradually removes layers of rock.

- Some organisms, for example sponges, molluscs and sea urchins, bore into rock. Coasts made up of limestone are particularly affected in this way.
- The mechanical action of plant roots can break open rocks, allowing other destructive processes to become effective.
- Some species of birds and mammals, for example puffins and rabbits, burrow into cliff faces, making weathering and erosion more likely.

Shoreline weathering: sub-aerial weathering and mass movement

The climate and weather of the coastal zone are important inputs to the coastal system. The same range of weathering processes that occurs inland may be active at the coast. However, the presence of seawater and the effect of tides in wetting and drying rock bring additional destructive influences. There is also the effect of past climates — for example, the changes resulting from the coming and going of ice.

Sub-aerial weathering

A range of weathering processes can be active on a stretch of coastline. However, the type of rock making up the coast, together with the climate, influences the processes that take place.

Mechanical weathering breaks off rock fragments of varying sizes, which fall to the foot of the cliff. Here they influence the rates of erosion of the cliff and the shore platform (Table 2.1).

Chemical weathering acts to decompose rock by altering the minerals within the rock (Table 2.2).

Table 2.1 **Effects of physical/mechanical weathering processes on coasts**

Weathering process	Effect
Crystal growth	Expansion of crystals (e.g. salt) when seawater collects in cracks in the cliff face and evaporates; as crystals grow, pressure is exerted on the rock
Freeze–thaw	Type of crystal growth involving repeated freezing and thawing of water; most effective on high-latitude coasts with significant precipitation
Wetting and drying	Expansion and contraction of minerals; most effective on clay

Table 2.2 **Effects of chemical weathering processes on coasts**

Weathering process	Effect
Solution	Solubility of minerals depends on temperature and acidity of the water; limestones are affected by carbonation, although they are not very soluble in seawater
Hydration	Minerals absorb water, weakening their crystal structure; rock is then more susceptible to other weathering processes
Hydrolysis	Reaction between mineral and water related to hydrogen ion concentration in water; particularly affects feldspar minerals in granite
Oxidation/reduction	Adding or removing oxygen; oxidation results from oxygen dissolved in water and particularly affects rocks with high iron content; reduction is common under waterlogged conditions
Chelation	Organic acids, produced by plant roots and decaying organic matter, bind to metal ions, causing weathering

- The cycle of wetting and drying as a result of tides has a significant effect on the coastal system. The set of processes most effective under such conditions are known as **water-layer weathering.** The zone of the coast affected extends between the low and high tide marks and into where spray is thrown. As a result, tidal range and meteorological factors, such as air temperature, are important — for example, in dictating how quickly evaporation takes place.
- Salt-crystal growth is probably the most significant weathering process.
- Corrosion of minerals, such as the solution of limestone, is common.

Activity 2

Refer to Figure 2.2 in Activity 1. For each rock type, describe and explain which sub-aerial weathering processes might be operating.

Mass movement

The coastal system has some of the most dramatic slopes in the world, in the form of towering cliffs. There are also gentle slopes at the coast. The full range of slopes in the coastal zone is subject to the downslope movement of material under the influence of gravity, which is called **mass movement**. The geology of the coast is an important influence on the type of mass movement. Many mass movements are triggered by undercutting from wave action at the base of a coastal cliff.

There are four types of rapid mass movement:

- **Rockfall** Blocks of rock dislodged by weathering fall to the cliff foot (e.g. granite cliffs, Costa Brava, Spain, see Figure 2.3).
- **Rock slide** Blocks of rock slide down the cliff face along seaward-dipping bedding planes (e.g. limestone cliffs, south Wales, see Figure 2.4).

| Figure 2.3 | **Rock fall from granite cliffs, Costa Brava, Spain** | Figure 2.4 | **Seaward dipping limestone strata likely to undergo rockslide, south Wales** |

Michael Raw

Tom Stiff

- **Rock toppling** Blocks, or even columns of rock, weakened by weathering fall seawards (e.g. sandstone and shale cliffs in north Devon).
- **Rotational slides and slumps** In a rotational slide, sections of the cliff give way along a well defined concave slip surface. The fallen material stays as an identifiable mass until further weathering and erosion act on it. Slumps occur

when a section of cliff collapses as a jumbled mass of rock. Slumps are common where permeable rock lies over impermeable rock (e.g. Christchurch Bay, southern England) or where the cliff consists of unconsolidated rock (e.g. glacial deposits in north Norfolk and Holderness, both in eastern England). There are two types of slow movement:

- **Creep** is the extremely slow downslope movement of regolith. Regolith is the layer of loose material, including soil, above bedrock. Any geology can be affected by creep.
- **Solifluction** is the slow downslope movement of waterlogged regolith. It is common at the end of an ice age when the surface layer begins to thaw, and also in summer in periglacial climates. It is active on high-latitude coasts but there is evidence for it having occurred in the past on some mid-latitude coasts (e.g. Somerset, western England).

The balance between marine erosion and sub-aerial weathering and mass movements is an important influence on cliff profiles and cliff recession.

Landforms

A rocky coast has a rugged morphology that has developed over a relatively long geological time scale. Increasing attention, however, is being paid to the effect of extreme events, such as tsunamis and 1 in 500-year storms, in shaping such coastlines. At any particular location, a combination of factors influences landforms.

Coasts in plan

When viewed from above, the shape of the coastline varies from one place to another. It can be a relatively straight line or be made up of bays and headlands. This variation is clear at a variety of scales.

Activity 3

Choose a stretch of coastline about 30–40 km in length.

(a) Trace the selected coastline from an atlas map at a scale of about 1:1 000 000.

(b) Trace part of the selected coastline from an OS map at a scale of 1:50 000.

(c) Trace the same part of the selected coastline from an OS map at a scale of 1:25 000.

(d) Comment on the influence of scale on the plan of the coastline.

Variations in plan are influenced by geology. **Lithology** and **structure** affect the way weathering and erosion operate. The lithology of a cliff describes the mechanical and chemical properties of the rock. Structure refers to factors such as the angle of dip of sedimentary strata, the extent of folding or faulting and the presence of intrusions such as bands of igneous rock. The direction in which joints lie is also significant.

Differential denudation is the process that causes some rocks to be worn away more rapidly than others.

Medium-scale to large-scale coastal plan

Where the coastline has a relatively straight plan with no large bays and headlands, it is known as a **concordant** coastline. Examples are the southwest and north stretches of the Gower Peninsula, south Wales (Figure 2.5).

| Figure 2.5 | **The geology of the Gower Peninsula, south Wales** |

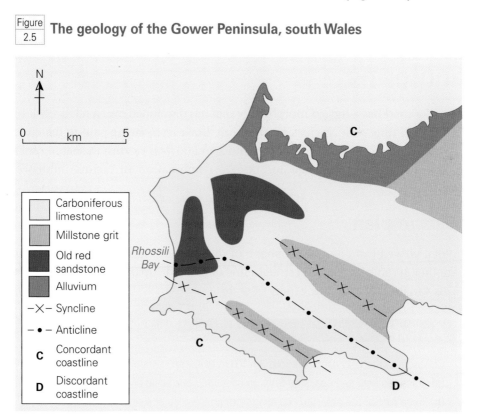

If the rock forming the cliff-line is particularly resistant, coves (small bays) and bays can form once a breach has been made. Either the sea or rivers can make this opening. Marine and sub-aerial processes can then attack the less resistant rock just inland (Figure 2.6).

| Figure 2.6 | **Concordant coastline, southwest Gower Peninsula, south Wales, showing coves in carboniferous limestone** |

At locations where the main rock types are at right angles to the coast, significant bays and headlands develop. This type of plan is known as a **discordant coastline**. The southeastern part of the Gower Peninsula is an example. The more resistant rocks form the headlands, with the bays carved into the less resistant geology (Figure 2.5).

It is important to appreciate the relief of the land immediately adjacent to the coastline, which is influenced by its geology. Folded structures produce anticlines and synclines. These can lead to the formation of ridges and valleys. Where these structures meet the coast, headlands form from the ridges and bays result from the valleys (Figure 2.5).

Patterns of wave energy vary between concordant and discordant coastlines. These variations affect where erosion and deposition occur and, therefore, influence the development of the coast.

Activity 4

(a) Trace the outline of the Gower Peninsula from Figure 2.5.

(b) Plot a series of approaching wave crests from the southwest, i.e. the direction of the prevailing winds. As each successive wave crest is drawn closer to the shore, use the principles of wave refraction to distort the line.

(c) Mark on the pattern of wave rays.

(d) Describe the pattern of high energy and low energy along this coastline. Suggest how this might influence:
- the processes occurring along the coast
- the development of the coast

Small-scale coastal plan

When looking in detail at a stretch of coastline, small-scale features become clear. In some locations these can be due to areas of relative weakness within a single rock type. For example, joints, bedding planes and faults are where rock is prone to weathering and erosion processes, in particular mechanical wave action. Depending on the precise arrangement of areas of weakness, small-scale landforms can develop. A **geo** is a narrow cleft that follows a line of weakness inland. It may have begun as a long and narrow cave but then its roof collapsed (Figure 2.7).

| Figure 2.7 | Geo following a fault line in Torridonian sandstone, Handa Island, Assynt, north Scotland |

Michael Raw

Small-scale coves and headlands are likely to owe their origin, in part, to differential erosion.

Marine cliffs: plan and profile

Steep slopes — cliffs — along the coastline, vary in both their plan and profile (cross-section). The cliff system helps us to understand why the morphology (shape) of cliffs varies from one place to another (Figure 2.8).

<table>
<tr><td>Figure
2.8</td><td>**The cliff system**</td></tr>
</table>

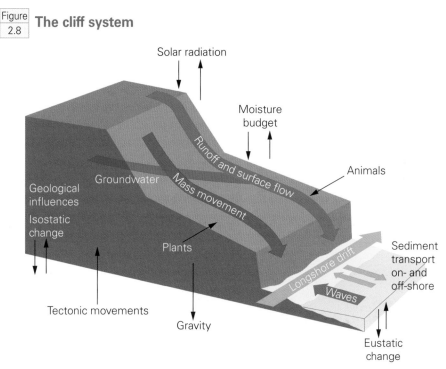

Different inputs result in different cliff morphology.

The role of geology

Most igneous and metamorphic rocks have lithologies resistant enough to form steep slopes (Figure 2.9).

Some sedimentary rocks, for example carboniferous limestone, old red sandstone and chalk, also form steep cliffs.

In contrast, unconsolidated rocks such as clays and recently deposited sands tend to result in low-angled cliffs. However, local conditions can result in steep cliffs in relatively weak rock. Constant marine erosion at the cliff base leads to both slope failure and mass movement of material onto the beach. Clays and marl, sands and glacial till are soon broken down and transported by waves and currents, leaving a steep slope to be undercut again (Figure 2.10).

Steep cliffs are often associated with either horizontal or vertical geological structure.

Figure 2.9 Castellated cliffs in granite, Jersey

Figure 2.10 Slumping in marl cliffs, east Devon

Peter Stiff

Peter Stiff

These slopes tend to retreat parallel to their faces as undercutting leads to rockfall and toppling (Figure 2.11(a) and (b)).

Cliffs comprising seaward dipping strata (Figure 2.11(c)) tend to maintain an angle similar to that of the dip as loosened blocks slide down to the cliff foot (see Figure 2.4, p. 28).

Where the strata dip landwards (Figure 2.11(d)) the profile tends to be relatively stable. Often, such slopes are slightly convex because marine processes are less effective than the sub-aerial attack on the upper portion of the cliff. The average angle of the joint pattern also influences cliff shape (Figure 2.11(e)).

At locations where cliffs contain more than one rock type, profiles are more complex. One characteristic profile found in areas where either ice sheets left till sheets behind or where periglacial processes were active is known as slope-over-wall (Figure 2.11(f)). A steep lower portion has a more gentle upper part. Active marine erosion creates the lower profile while sub-aerial processes, both present-day and those operating in the past, result in upper-profile slopes of about 25–30°.

Figure 2.11 **Cliff profiles: the influence of lithology**

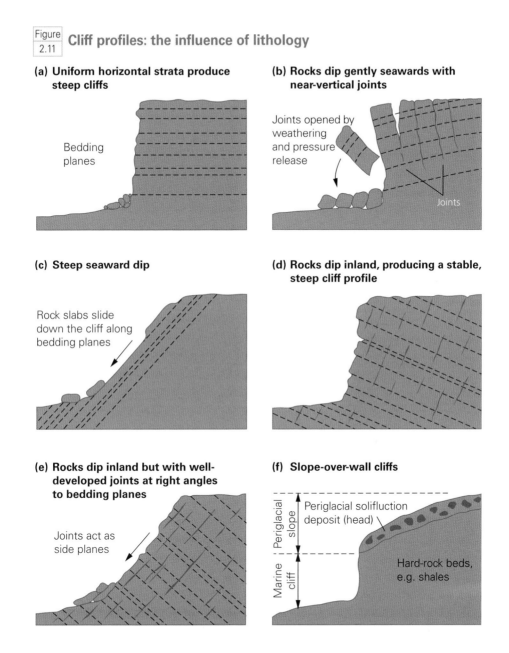

(a) Uniform horizontal strata produce steep cliffs

Bedding planes

(b) Rocks dip gently seawards with near-vertical joints

Joints opened by weathering and pressure release

Joints

(c) Steep seaward dip

Rock slabs slide down the cliff along bedding planes

(d) Rocks dip inland, producing a stable, steep cliff profile

(e) Rocks dip inland but with well-developed joints at right angles to bedding planes

Joints act as side planes

(f) Slope-over-wall cliffs

Periglacial slope

Marine cliff

Periglacial solifluction deposit (head)

Hard-rock beds, e.g. shales

The balance between marine and sub-aerial processes

Marine processes are important in two key areas of cliff development. Marine erosion attacks the cliff between low and high tide levels. Debris from the eroded cliff, as well as material produced by sub-aerial processes operating higher up the cliff face, falls to the cliff base. Marine processes then remove this sediment.

Activity 5

Assume a cliff profile as in Figure 2.12.

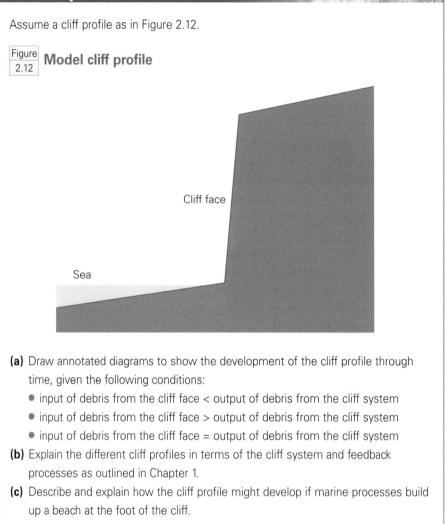

Figure 2.12 **Model cliff profile**

(a) Draw annotated diagrams to show the development of the cliff profile through time, given the following conditions:
- input of debris from the cliff face < output of debris from the cliff system
- input of debris from the cliff face > output of debris from the cliff system
- input of debris from the cliff face = output of debris from the cliff system

(b) Explain the different cliff profiles in terms of the cliff system and feedback processes as outlined in Chapter 1.

(c) Describe and explain how the cliff profile might develop if marine processes build up a beach at the foot of the cliff.

It has been suggested that cliff form varies with latitude because the relative rates of debris supply and removal are related to the coastal energy system. Wave energy is highest along mid-latitude coasts and lowest in the low latitudes (tropics). Areas of high latitude tend to be either relatively sheltered or experience ice in the winter and, therefore, receive low wave energy. Debris removal and supply from marine processes are at a maximum in the mid-latitudes.

Variations in sediment yield from sub-aerial sources are more complex. Humid, low-latitude areas tend to have slopes covered by vegetation, which yield relatively small amounts of sediment. However, arid tropical coasts lack vegetation cover and so steep cliffs are more common. Weathering processes such as frost shattering are active in high-latitude locations. Large volumes of scree are produced, which low wave energy is incapable of removing. Here, the slopes reflect the angle of rest of the sediment that makes up the scree.

Features of cliff retreat: plan and profile

Where there is a clear divide between the sea and the land, for example along coastlines in which the land surface is significantly above sea level, a variety of distinctive landforms can develop.

A **notch** that extends underneath the cliff face can develop around the mean high water mark. Notches tend to result from a combination of wave erosion, bioerosion and weathering. They are found in different geologies and are particularly common in limestone.

Sea caves develop around the mean water level and extend into the cliff base. They can measure 50 m or more from front to back. Structural weaknesses in rock, for example joints and bedding planes, are exploited by wave action. Where the rock has vertical lines of weakness, a **blowhole** can be created, extending through the roof of the cave and opening onto the cliff top. When the tide is high enough, water can be forced through this natural pipe to give a spectacular release of water at the cliff top. Spouting Horn on Kaua'i, one of the Hawaiian chain of islands, is the result of differential erosion of a lava tube. It regularly produces a fountain 15 m high (Figure 2.13).

At locations where small-scale headlands protrude from the coastline, wave refraction focuses energy onto the headlands. Notches and caves can extend through the headland to give an **arch**. These features can last for decades, but eventually they collapse to leave an isolated column of rock — a **stack**. The arch that was part of Marsden Rock collapsed in 1996 (Figure 2.14(a) and (b)).

Figure 2.13 **Spouting Horn, Kaua'i Island**

Peter Stiff

Figure 2.14 Marsden Rock, Tyne and Wear, 1981 (a) and after its collapse in 1996 (b)

(a)

Michael Raw

(b)

Michael Raw

Stacks continue to wear away until a **stump** (a small platform of rock) is left, which is sometimes covered at high tide (Figure 2.15).

Shore platforms

Shore platforms are horizontal or gently sloping landforms extending seawards from cliffs. They are relatively flat rock surfaces, with rock pools and occasional larger blocks of rock. High tides tend to cover them; they are exposed at low tide (Figure 2.16).

It is clear that shore platforms are intrinsically linked with cliff retreat. However, it would be incorrect to assume that the same processes created both cliffs and platforms.

Figure 2.15 **Stacks and stumps, Bedruthan, Cornwall**

Michael Morrish

Figure 2.16 **Shore platform in resistant Jurassic sandstone and ironstone, Saltwick Bay, North Yorkshire**

Michael Raw

Wave-cut platform was the term used to indicate the assumed predominance of wave action in the production of this landform. However, it is now recognised that a number of processes are responsible and, also, that in various parts of the world, the significance of different processes varies. Shore platforms in sheltered locations are unlikely to have been carved solely by wave action. Where wave energy is high, the role of weathering in partially weakening rock before the waves complete the wearing away should not be underestimated. In particular, wetting and drying and salt weathering are effective on shore platforms. In some locations, bioerosion plays an important role. The debate regarding the relative importance of these processes in the formation of platforms continues.

Two basic types of shore platform tend to be recognised:

- sloping — gradient 1–5°, a continuous platform without major interruptions in the slope
- sub-horizontal — gradient almost flat with a small rampart and low cliff at the low water mark (Figure 2.17)

Figure 2.17	**Shore platforms**

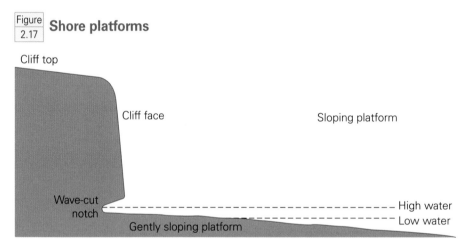

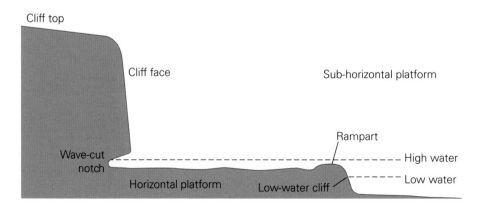

Some geographers suggest that sloping platforms are common in macro-tidal environments such as the British Isles. The sub-horizontal are more likely to occur along coasts with meso- and micro-tidal regimes — for example, most of Australia and the Mediterranean.

Activity 6

Assume a cliff profile as in Figure 2.12 (p. 36).

Figure 2.18 **The relationship between shore platform development, wave energy and cliff retreat**

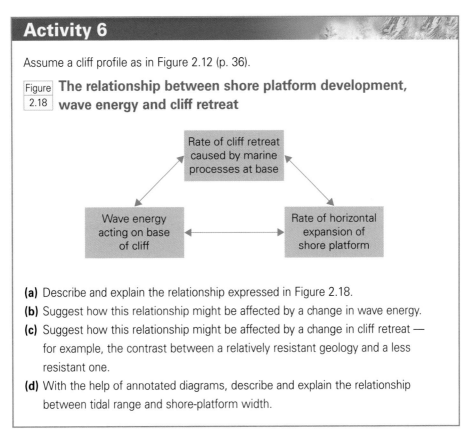

(a) Describe and explain the relationship expressed in Figure 2.18.

(b) Suggest how this relationship might be affected by a change in wave energy.

(c) Suggest how this relationship might be affected by a change in cliff retreat — for example, the contrast between a relatively resistant geology and a less resistant one.

(d) With the help of annotated diagrams, describe and explain the relationship between tidal range and shore-platform width.

3 Beach environments

Sediment produced by erosion, weathering and mass movement is an important element of the coastal system. Sediments are transported by waves and currents and are deposited elsewhere to make new landforms.

Sources of coastal sediments

There is a wide range of shapes, sizes and mineral make-ups in sediments found in the coastal zone. This reflects the diversity of the materials broken down to form the sediment. Although the obvious sources of coastal sediments are the coastal landforms, for example cliffs, 90% of coastal sediment comes from the denudation of areas inland of the coast and is transported to the sea by rivers (Figure 3.1).

Figure 3.1	**Coastal sediment budget**

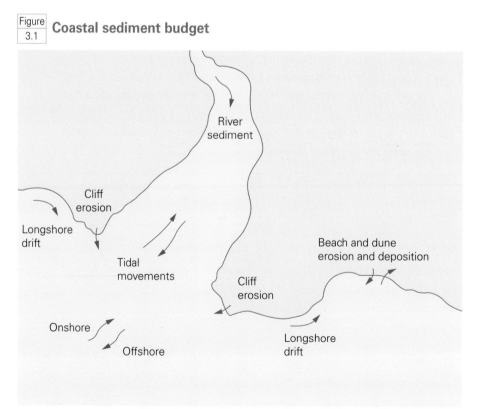

Where wave energy is low, especially in the tropics, river sediments are particularly important. Rivers such as the Niger and the Amazon carry high levels of suspended loads into the coastal zone.

There are, however, stretches of coastline where local geology plays an important role in supplying sediment. For example, Black Sand Beach at Punalu'u, Hawaii reflects its volcanic origins. Along stretches of the southern Californian coast, recent research shows that as much as 50% of the beach sand has come from the collapse of relatively weak sedimentary cliffs. Beaches and dune systems can erode, with their sediments being carried from the land to the sea. The production of some sediments, for example shell, is generated entirely within the coastal zone.

Sediments can also be transported onshore. These sediments might be former beach or river material that initially had been carried offshore. Exceptionally energetic waves can bring sediment onshore but most onshore movement is due to the rise in sea level following the last ice age. This is an important point. It reminds us that landforms cannot always be understood by considering only the processes operating today.

Types of sediment

There are two principal groups of sediment type:

- **Clastic** sediments are broken down rocks. These can be:
  - **lithogenic** — fragments of rock (e.g. pebbles, see Figure 3.2)
  - **minerogenic** — rock broken down into individual mineral grains (e.g. quartz)
- **Biogenic** sediments are the remains of materials such as corals and shells.

Sediments are also classified in terms of size and shape. Size is commonly taken as the diameter of the particle. There is such a wide range of sediment sizes in the coastal zone that a logarithmic scale is used. This accounts for the categories in Table 3.1. It is important to appreciate the enormous range in size — for example, between clay/silt and coarse sand grains of diameter 1mm. Very coarse sand is 1000 times larger than clay particles.

Figure 3.2 **Pebble/shingle mixture**

Peter Stiff

| Table 3.1 | **Particle names and sizes** |

Table 3.1 Particle names and sizes

Particle name	Relative size	Actual size (mm)
Boulders	Very large	2048
	Large	1024
	Medium	512
	Small	256
Cobbles	Large	128
	Small	64
Pebbles/shingle	Coarse	32
	Medium	16
	Medium	8
	Fine	4
Sand	Very coarse	2
	Coarse	1
	Medium	0.5
	Medium	0.25
	Fine	0.125
	Very fine	0.063
Silt	Very coarse	0.031
	Coarse	0.016
	Medium	0.008
	Fine	0.004
Clay	–	0.002

When studying sediment that is pebble-sized or larger, the three axes of an individual particle are used to help describe shape (Figure 3.3).

Figure 3.3 The axes of a particle

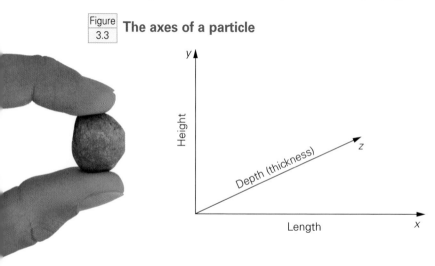

Using the measurements of the three axes, particles of the same size can be distinguished from one another. The four main categories of shape are:
- rod — long and thin
- sphere — ball-like
- blade — long and flat
- disc — round and flat

Processes

When there is sufficient energy, sediment is picked up and transported by a number of processes. Deposition occurs when and where there is insufficient energy to move sediment further (Figure 3.4).

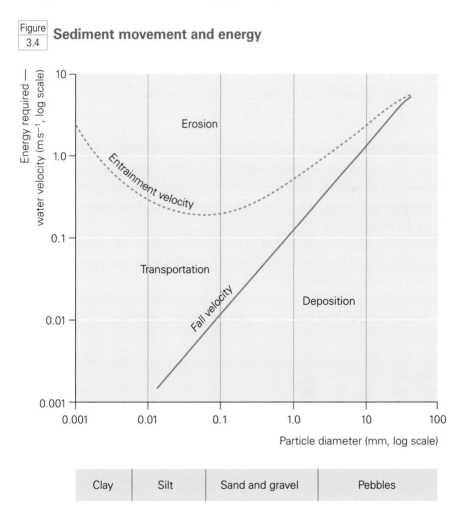

Figure 3.4 **Sediment movement and energy**

Sediment movement

In the coastal zone, sediment often moves in water. This is known as the **load** of the water. Winds, and occasionally ice, are also agents that can move particles in the coastal zone.

In general, the larger the size of particle, the more energy is required to pick up and entrain it. However, the smallest particles — silt and clay — are exceptions. These are electrically bonded together, which makes them more cohesive. Therefore, more energy is required to lift them off the seabed than is required to lift sand grains. Silt and clay particles also present a smoother surface for the water to push against compared with sand grains. Once these small particles are entrained, little energy is required to keep them moving. Larger sediment particles, such as pebbles and cobbles, are soon deposited once moving water begins to slow. These relationships are summarised in Figure 3.4.

Breaking waves cause turbulence, which makes sediment transport in the coastal zone a complex process. Water movement is not just in one direction. Swash and backwash carry sediment up and down a beach. Water can move along a shore and flow back out to sea, via rip currents, carrying sediment with it.

Types of sediment movement

Larger sediment particles roll, slide and hop along the seabed in a process called **traction**. This movement is intermittent because of the variation in energy input via waves. It is possible to hear pebbles being moved on a shingle beach when high-energy waves break. When the same beach receives low-energy waves, few pebbles are moved.

Saltation is a skipping movement of sand grains that occurs along the seabed or on a dry beach when the current flows or the wind blows with enough energy to entrain particles (Figure 3.5).

Figure 3.5 Saltation and the transport of sand-sized particles

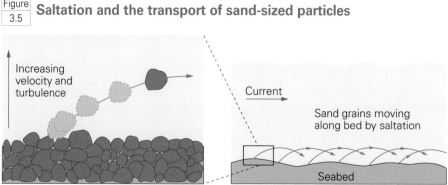

Individual sand grains move along the surface in an arc-shaped trajectory. When an individual grain lands, it disturbs more grains, causing these to be transported. Relatively short distances are travelled in any one 'skip' by a sand grain. However, the process is cumulative and vast quantities of sand can be moved. This is a key process in the evolution of sand dunes.

When there is plenty of energy in moving water, **suspension** occurs. Sediment, usually sand grains and finer particles, is kept moving by turbulent water. Individual grains are buffeted by eddies, with the finer clays and silts kept moving in all but low-energy conditions. There is no clear division between suspended load and bedload (traction and saltation combined). Measurements in the coastal zone are difficult to make and there is the added complication of vertical differences in energy within the water.

Minerals are transported in **solution**. Limestone rocks such as chalk can be weathered and dissolved by seawater with the solute carried by the sea.

Onshore movement of sediment

Below wave-base (p. 10), water is not moved by passing waves. At this depth, therefore, sediment is not disturbed and so cannot be transported by wave action. There are, however, currents of water movement that can move the sediment. Orthogonal movement of sediment occurs perpendicular to the coasts. Sediments may be moved offshore and onshore. For example, in Christchurch Bay, Hampshire, shingle is moved onshore from the shingle bank to some of the beaches along the bay.

The rise in sea level (100–200 m) that followed the last ice age transported vast quantities of sediment. As sea level rose, water moved across extensive areas covered by glacial and periglacial debris. This material was entrained, altered in shape and size by attrition and, when sea-level rise stabilised, deposited. There are landforms built from significant accumulations of sediment around coastlines in the mid-latitudes, the presence of which can only be explained by these processes. Chesil Beach in southern England is one such landform.

Offshore movement of sediment

- Material can be moved offshore during storm conditions. High-energy waves are capable of transporting much material in just one storm. A severe storm coinciding with high tide cannot only lower a beach but can also remove unconsolidated material, such as sand dunes, from inland.
- Material moved along a shore might reach a river estuary and be carried offshore by the current of water discharging from the river out to sea.

- The seabed is not uniformly flat. A relatively common feature is a submarine canyon extending offshore. Water flows down and along the canyon carrying with it sediment that is, therefore, removed from the shore.

Longshore movement of sediment

When the angle of wave approach to the shore is oblique, the advancing swash moves up the beach at the same angle. Sediment is entrained by the moving water and transported up the beach in the swash. The wave eventually runs out of energy, because of friction and the gradient of the beach profile. Some sediment is likely to be deposited at the point where the swash finally ceases. Smaller particles might continue to move with water that returns to the sea. This water returns to the sea directly under the influence of gravity, which acts perpendicular to the beach slope. The next wave then breaks and the process is repeated. Sediment is once again entrained and transported longshore in the direction of wave approach (Figure 3.6).

| Figure 3.6 | **Longshore drift** |

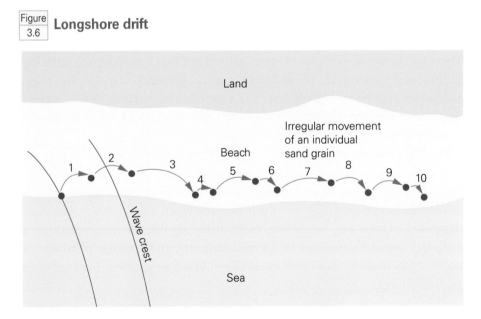

Longshore drift is an irregular process. No two waves are identical and as a result sediment is moved in slightly different paths. The angle of wave approach may have a dominant direction but there will be times when the wind and, therefore, the waves come onto the beach from a different direction. Therefore, longshore movement is not a simple process. In the short term (on a day to day basis) it is variable. In the longer term, over months and years, it operates in a preferred direction.

Activity 1

Criccieth beach is a shingle accumulation on the northern shore of Cardigan Bay, northwest Wales (Figure 3.7).

Figure 3.7 **Location of Criccieth beach**

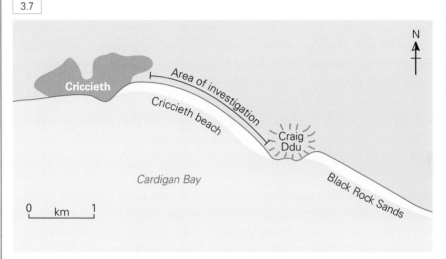

An investigation was carried out in order to answer the question, 'How and why does the size of sediment vary along the beach?' Samples of shingle were taken at ten locations spread west to east along the beach at intervals of 25 m.

Table 3.2

Sampling point	1 (W)	2	3	4	5	6	7	8	9	10 (E)
Particles <10 mm x-axis (%)	3	4	12	16	9	26	42	34	60	73

(a) Use the data in Table 3.2 to draw a scattergraph to show the relationship between distance (west to east) along the beach and the percentage of particles with x-axis less than 10 mm.

(b) Analyse the statistical significance of the relationship between distance along the beach (west to east) and sediment size using a correlation coefficient.

(c) Comment on the result and its significance for changes in sediment size.

(d) Evaluate the factors likely to be involved in producing the pattern of sediment along Criccieth beach.

Landforms

The landforms created by sediments can be just as spectacular as those found along rocky coasts. However, they also have subtlety in both their formative processes and shapes. The most widespread category of depositional landform is the beach.

Beaches are stores of loose sediment within the coastal system. A variety of unconsolidated material, most commonly sand or shingle, can accumulate between the area where waves begin to experience friction with the seabed to the zone landwards of the high tide level. Beaches experience almost continual changes in shape. Sand, shingle and pebbles respond rapidly to changing inputs of energy from waves, tides and winds. Many beaches exist in a state of dynamic equilibrium. There are also occasions when dramatic changes occur, particularly in response to abrupt increases in wave energy, for example during a high-energy storm.

In the mid and high latitudes, pebble beaches are most common; in the low latitudes, sand beaches prevail. This spatial contrast might be due to the sediment input in the low latitudes being dominated by fine, fluvial sediments. In the higher latitudes, larger-calibre sediment is common because of the significant input of material by glacial and periglacial processes.

Beach shape can be studied in two directions. Seen from above, as a bird's eye view, is the beach **plan**. The cross section of a beach is its **profile**. As with many landforms, beaches can be investigated at more than one scale, ranging from about 1m to more than 100km. Within the larger (macro) scale, such as a whole beach, exist smaller (micro) scale landforms such as **cusps** and **berms**.

Beach plan: large-scale landforms

Beaches tend to follow the general trend of the coast. However, some extend out from the coastline. The main influence on beach plan is wave energy — in particular, the relationship with the prevailing wave direction. There are two fundamentally different types of beach plan:

- **swash-aligned** beach — oriented parallel to the shoreline
- **drift-aligned** beach — oriented obliquely to the shoreline

A swash-aligned beach is usually a fairly closed sediment system as there is limited transfer of material into or out of the beach (Figure 3.8). Wave refraction causes wave crests to curve to the shape of the shore. The angle of wave approach reduces to a point when longshore transport of sediment is minimal and deposition predominates.

Figure 3.8 **Crescent shaped swash-aligned beach plan, Rhossili Bay, Gower, south Wales**

Tom Stiff

A drift-aligned beach is a more open system. Sediment enters at one end, passes along the length of the beach due to longshore drift and then leaves the beach at the far end. The angle of wave approach is great enough to keep sediment moving.

Within a relatively short stretch of coast, a few tens of kilometres, both swash- and drift-aligned beaches can be found (Figure 3.9).

Figure 3.9 **Beaches on part of the Lleyn Peninsula, north Wales**

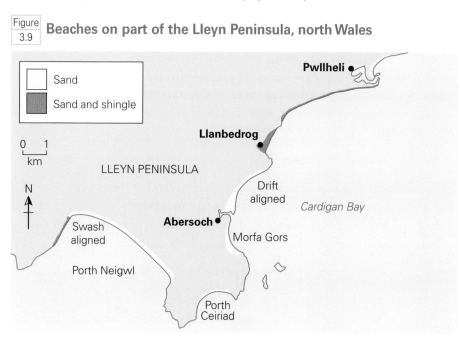

The more enclosed a bay is, the more refracted the waves tend to be. This process results in **bay-head** or **pocket** beaches. There is one type of beach plan that seems to be the result of both swash and drift processes. Where headlands interrupt longshore drift only partially, **zeta-form** or **fish-hook** beaches develop (Figure 3.10).

Figure 3.10	**Zeta-form beach**

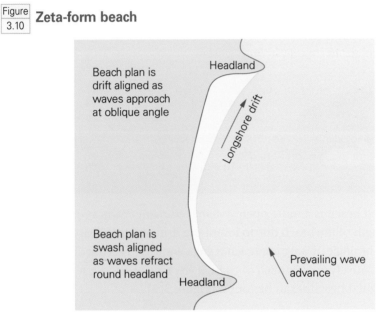

Refraction around the headland produces wave crests that have a small angle of approach to the shore. Further along the bay the angle of wave approach is greater, which allows longshore transport of sediment.

Beaches that extend out from the general direction of the coastline are known as **detached** beaches. However, some remain in contact with the coast for part of their length.

Spits

Spits are linear deposits of sediment attached to land at the proximal end but free at the distal end. They are found at locations where:
- the coast has an abrupt change in direction, such as at an estuary or bay
- there is a ready supply of sediment, particularly sand and shingle
- longshore drift is active
- tidal range is limited (less than 3m)

A beach accumulates out from the shore in the direction that longshore drift is transporting sediment. The tidal range is narrow, so wave energy is focused into a restricted zone and transports and shapes drift-aligned features.

Where the longshore current enters deeper water near the tip of the spit, energy is dispersed throughout more water. Rates of transport are reduced and deposition increases. In addition, waves refract around the end of the spit, taking sediment with them. In the relatively low-energy conditions just behind the distal end of the spit, sediment builds up in the form of a curved ridge. Some spits have a number of **recurves** (hooks or laterals) along their length (Figure 3.11). It is likely that where the recurve is pronounced, waves, approaching from a different direction to the main direction of drift, bring enough energy into the spit system to move sediment in that different direction.

Figure 3.11 **Recurves on Hurst Castle spit, Hampshire**

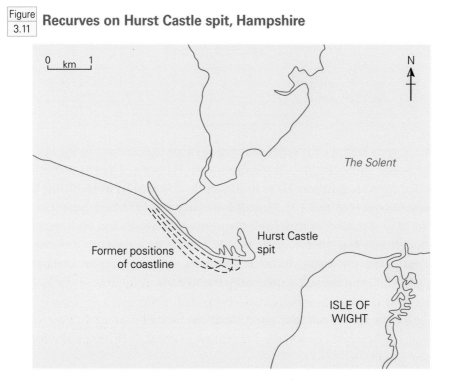

Recurves can mark different phases in the development of a spit. They can be associated with periods when the coastline, including the spit, was seaward of its later position.

Most spits have formed since sea level stabilised about 4000 years ago and there is good evidence to indicate that many are much younger. Spits are landforms that adjust relatively quickly to changing inputs of energy and sediment. In the course of a couple of centuries, a spit can extend across the mouth of a river to divert its exit to the sea by several kilometres. For example, in Suffolk, the exit of the River Alde into the North Sea has been diverted south by some 12 km due to the growth of Orford Ness spit (Figure 3.12).

Figure
3.12 **River mouth diversion due to spit growth**

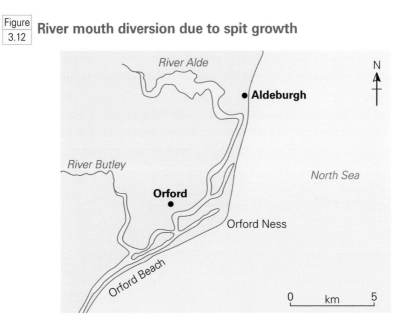

A pair of spits facing each other either side of an indentation in the coastline might seem contradictory. At Christchurch Harbour, Dorset, a spit once stretched from Hengistbury Head in the south almost as far as Highcliffe Castle to the northeast (Figure 3.13). Violent easterly storms breached the spit in 1886 and in 1935. Evidence of drift from the north is unconvincing, so it seems that the double spit was once a single feature formed by longshore movement of sediment from the south. In some locations, drift takes place in opposing directions for long enough to generate double spits.

Figure
3.13 **Double spit at Christchurch Harbour, Dorset**

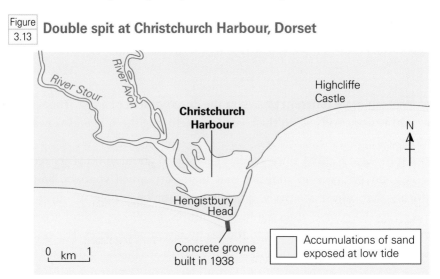

The very existence of a spit is often a precarious balance between inputs of wind, wave energy, tidal energy and sediment.

Activity 2

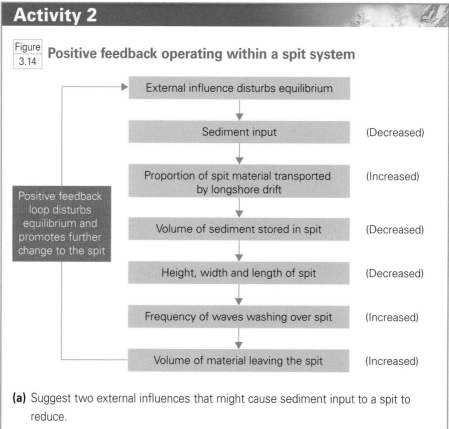

Figure 3.14   **Positive feedback operating within a spit system**

External influence disturbs equilibrium

Sediment input   (Decreased)

Proportion of spit material transported by longshore drift   (Increased)

Positive feedback loop disturbs equilibrium and promotes further change to the spit

Volume of sediment stored in spit   (Decreased)

Height, width and length of spit   (Decreased)

Frequency of waves washing over spit   (Increased)

Volume of material leaving the spit   (Increased)

(a) Suggest two external influences that might cause sediment input to a spit to reduce.

(b) Redraw Figure 3.14 to show the changes that might occur if sediment input increased. Explain the circumstances under which this might happen.

(c) Redraw Figure 3.14 to show the changes that might occur if wave energy increased.

Activity 3

Use an atlas map to set the part of the northwest Wales coast shown in Figure 3.9 (p. 51) in a wider context. In particular, note the contrasting aspects of the various segments of this stretch of coastline.

(a) Describe the different types of beaches and their distribution along this stretch of coastline.

(b) Using annotated diagrams and sketch maps, suggest how the beaches result from interaction between different factors.

Cuspate forelands and tombolos

A **cuspate foreland** is a triangular-shaped projection with its apex pointing out to sea. Cuspate forelands vary in scale. Dungeness foreland extends some 30 km along the Kent coast and projects about 15 km into the English Channel (Figure 3.15). Along the Carolina coast, USA, cuspate forelands can reach 150–200 km along the side attached to the mainland. An example is Cape Fear (see Figure 3.16, p. 58).

Figure | **The formation of Dungeness**
3.15

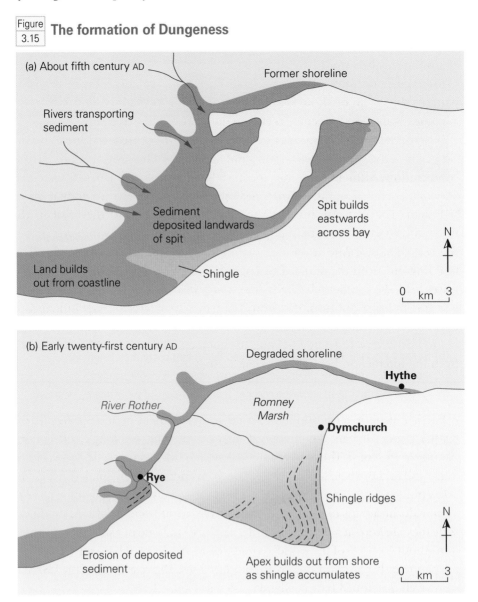

(a) About fifth century AD — Former shoreline

Rivers transporting sediment

Sediment deposited landwards of spit

Spit builds eastwards across bay

N

Land builds out from coastline — Shingle

0 km 3

(b) Early twenty-first century AD — Degraded shoreline

Hythe

River Rother — Romney Marsh

● Dymchurch

● Rye

Shingle ridges

N

Erosion of deposited sediment

Apex builds out from shore as shingle accumulates

0 km 3

Cuspate forelands seem to represent locations where sediment, moved by longshore drift, becomes trapped when equilibrium between sediment inputs and the energy available to move it is reached. However, different explanations account for their formation in some locations. Where there is an offshore island, waves refract around it on either side. Wave energy thus travels along the shore in opposite directions, meeting in the shadow of the island. Sediment is deposited in this area. At some locations, this process leads to a **tombolo** developing. This is a thin strip of sediment extending from an offshore island to the mainland. The town of Llandudno, north Wales, is built on a tombolo that links the mainland with the Great Orme, a former offshore island made of carboniferous limestone.

Some cuspate forelands are not associated with an alteration in the wave patterns due to an offshore island. The funnelling of tidal currents is relevant at some locations. Dungeness seems to have existed first as a spit.

The spit extended eastwards as longshore drift carried sediment in the direction of the westerly prevailing wind and waves. The combination of rivers contributing sediment from the land and the increasing influence, further east, of wave energy from the North Sea, probably aided the accumulation of sediment. The spit was breached where it was attached to the coast in the thirteenth century. From then on, sediment eroded from the southwest part of this stretch of coastline. It was deposited further east as a series of shingle ridges building out from the shore. The English Channel is narrow at this point, so waves from the southeast brought little energy into the coast. Therefore, the apex was able to build out from the shore. Human activity was another influence as drainage of marshland, changes in land use and coastal protection modified natural processes. Cuspate forelands are good examples of landforms resulting from the interaction between different factors.

Bars and barrier islands

A **bar** is a generic term applied to a range of sediment accumulations in the coastal zone. All bars are elongated deposits of sand or shingle, often lying parallel to the coast. They are submerged at high tide. A lagoon lies between the mainland and the bar. Bars range in scale from comparatively small features just a few metres wide and a couple of hundred metres long to landforms over 1km wide, hundreds of kilometres long and up to 100m in height. At this larger scale they are known as **barrier islands**. Barrier islands are not submerged at high tide. About 10–15% of the world's coastlines are made up of barrier islands. They are particularly common in low- to mid-latitudes. The east coast of the USA has some well-developed barrier islands systems (Figure 3.16).

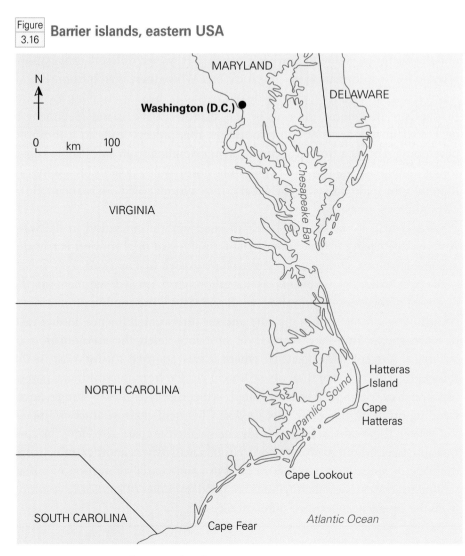

Figure 3.16 **Barrier islands, eastern USA**

Factors common to barrier islands are:
- a gently sloping offshore gradient
- limited tidal range (less than 3 m)
- relatively high wave energy

Most barrier islands have a series of dune ridges on their seaward face. On the landward side, washover fens and tidal marshes form. Large quantities of sediment are transported daily along and around barrier beaches. Where the barrier is made of coarser sediment such as shingle, seepage of water through the barrier can transport vast volumes of water into the lagoon. Tidal currents and the escape of river and lagoon water combine to produce significant fluctuations in the patterns of erosion and deposition along a barrier.

The role of changing sea level to bar and barrier-beach formation

The formation of barrier beaches continues to cause controversy; it is likely that at different locations and times, alternative theories apply. Cores obtained from barrier islands along the east coast of the USA show deposits of silt and clay beneath the sand or shingle. These small-calibre sediments are most likely to have formed in low-energy conditions such as estuaries or lagoons trapped behind the barrier. It seems that barriers rolled landwards as sea levels rose at the end of the last ice age (Figure 3.17).

Figure 3.17 **Barrier island formation**

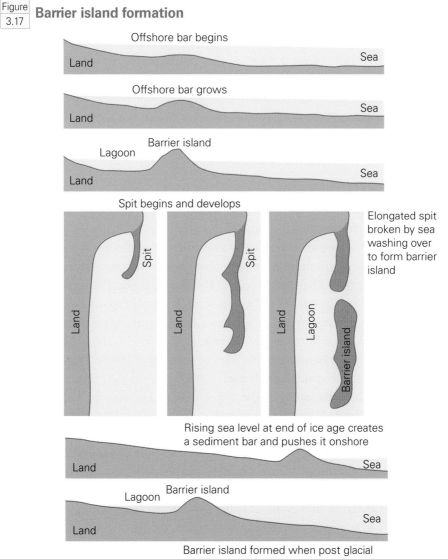

An alternative view suggests that the islands, originally sand dunes and beaches, were separated from the mainland by rising sea level as a result of glacio-eustasy (see p. 90). It has been suggested that, with a falling sea level, bars just offshore were exposed and became barrier islands. Some barriers are felt to be the result of former spits being breached by high-energy waves that then eroded a permanent channel to form islands (Figure 3.17).

The shingle accumulations around mid-latitude coasts such as the British Isles are thought to be more a product of past processes than of processes operating today. As a response to glacio-eustacy, relative sea level fell to 100–120 m lower than today. Therefore, vast areas of foreshore were left exposed. Intense periglacial processes, such as freeze–thaw weathering, operated to break down material at the surface. Then, sea levels rose when the ice age ended. Waves picked up the loose material and rolled it landwards. Angular sediment became rounded through attrition and eventually the growing bar of pebbles came to a halt against a cliff line. The largest shingle accumulation of this type is Chesil Beach. Shingle stretches for about 30km along the Dorset coast, trapping a lagoon behind it and linking the island of Portland with the mainland via a tombolo. Chesil beach reached its present position about 6000 years ago. It is a swash-aligned feature.

The equilibrium of barrier islands

As with spits, barrier islands are subject to considerable change as sea level rises. Sediment eroded from the seaward face is carried offshore. Alternatively, waves wash over the beach in storms. Under these high-energy conditions, sediment erodes from the seaward face of the beach and is carried over the barrier. It is then deposited in the lower energy area behind the barrier. The entire landform migrates landwards.

It is possible for the barrier to be overwhelmed by a rapidly rising sea level. Most sediment remains on the seabed as a relict feature, with a small quantity of material deposited around the higher, high tide level further inland.

It seems that many barrier beaches exist in a state of disequilibrium. With little sediment accumulating and wave energy increasing as sea level rises, the landform changes shape. In some parts of the world, where these islands are intensively used for human activity, for example the eastern seaboard of the USA and the northern Netherlands, the long-term survival of barrier beaches is a serious concern. Rising sea level is leading to increased erosion and transport of sediment. Increasing frequency and severity of storms results in waves over-topping barrier islands, threatening lives and property.

Offshore bars

Within the inter-tidal zone, nearshore bars can develop. These have various shapes and arrangements:

- linear and parallel to the shore
- crescent-shaped and parallel to the shore
- discontinuous
- linear, at an angle to the shore

The process of offshore bar formation is a matter of debate, but all theories stress the importance of wave energy. Where waves interact with the seabed at depths less than the wave base, material moves onshore and **breakpoint bars** form. In addition, material moves offshore, transported by backwash. A bar forms where the sediment flows converge. Nearshore bars can migrate both on and offshore. Onshore movement generally occurs when wave energy is low to moderate. Offshore migration is more often associated with higher energy waves, when swash flows are able to carry sediment along the bed. Higher energy waves tend to flatten beach profiles and result in a net loss of sediment to offshore (see Figure 3.21). Typical rates of migration lie in the range $1-10\,\mathrm{m\,day^{-1}}$, although rates as high as $30\,\mathrm{m\,day^{-1}}$ have been observed. There is also evidence of a cyclical pattern of migration, indicating dynamic equilibrium over a period of several years.

Ridge-and-runnel beaches

The combination of a sand beach, a limited fetch and a large tidal range often results in a ridge-and-runnel beach. A series of crests (ridges) and troughs (runnels) form (Figure 3.18). The distance between crests is 100–200 m. Vertical differences in height are small — about 1m from the bottom of a runnel to the top of a ridge. They tend to orientate parallel to the direction of maximum fetch. The ridges are not continuous; they are punctuated along their length by small channels. These channels allow water from the runnels to drain as the tide ebbs towards low tide.

| Figure 3.18 | **Runnels, Camel estuary, Cornwall** |

Michael Morrish

Beach plan: small-scale landforms

Within the overall plan of a beach, a number of smaller-scale features are found. Their small scale usually results in a short lifespan. Whereas a large beach or bar can persist for decades or centuries, small-scale features such as cusps, ridges and runnels may last for only a few days before their shapes are adjusted by changing inputs of energy and sediment.

Cusps are crescent-shaped features spaced regularly along a beach. They range in size from 1–60 m from one point of the crescent to the other. In general, the greater the input from wave energy, the larger are the cusps. The horns of the cusp consist of coarse sediment; finer sediments occur in the curve of the crescent (Figure 3.19).

| Figure 3.19 | **Water and sediment flows in cusps** |

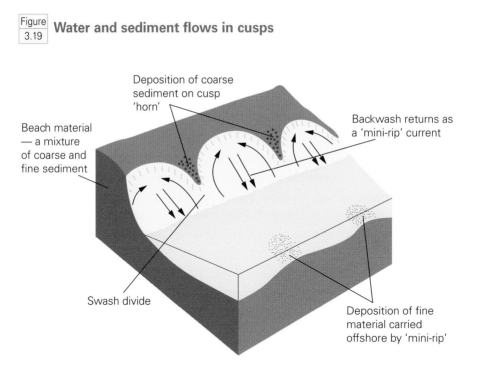

How cusps form is uncertain. However, once developed they reinforce the processes that form them. The model that receives the most support from fieldwork combines positive and negative feedback. This **self-organisation model** suggests that along a relatively straight beach face, water is attracted into small depressions and then accelerates. Sediment is eroded, which makes the depression deeper. Therefore, it attracts more water and the flow is further accelerated as positive feedback operates. Small ridges on the beach repel water and cause flow to decelerate. At these locations, lower energy results in deposition, which increases the size of the ridge. As the depressions become larger, negative feedback starts to operate as swash runs out of energy before reaching the back of the cusp and so no sediment is removed. Likewise, once the horns of the cusp have reached a certain size, water flows off quite quickly and sediment is not deposited.

Rip channels form on sand beaches where waves approach parallel or obliquely to the shore. Returning water escapes as rip currents that can scour channels in the sediment (see Figure 1.15, p. 19). The positions of rip channels shift in response to slight changes in the angle of wave approach.

The smallest features in a beach plan are **ripples** (Figure 3.20). They form in sand on a scale of up to 10 cm in height and up to 50 cm between crests. Symmetrical ripples form where water flows have similar velocities. Where there is a pronounced flow in one direction, asymmetric ripples are found. The strong current tends to encourage ripples to migrate.

| Figure 3.20 | **Ripples on sand beaches** |

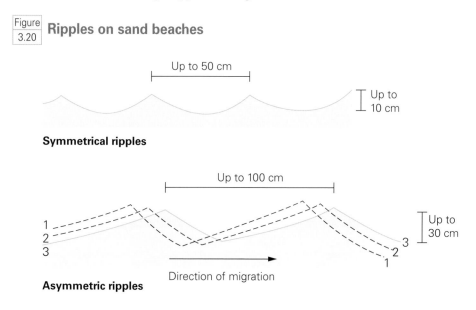

Symmetrical ripples

Up to 50 cm

Up to 10 cm

Asymmetric ripples

Up to 100 cm

Up to 30 cm

Direction of migration

Beach profile

As with any slope system, the beach profile reflects the interaction of a number of variables. Key factors are:

- wave energy
- size and shape of beach material
- tidal range

The role of wave energy

Laboratory experiments and fieldwork both suggest a strong relationship between waves (in particular, wave steepness) and the angle of the beach profile.

Activity 4

- Wave steepness is the ratio of wave height to wave length. The higher the value, the greater the energy brought by the wave onto the beach.
- The angle of beach profile is the mean angle of the beach face facing the sea.

The relationship between wave steepness and angle of beach profile for a sample of beaches is shown in Table 3.3.

Table 3.3 **The relationship between wave steepness and beach profile**

Beach	Wave steepness	Angle of beach profile
1	0.24	8
2	0.225	10
3	0.22	9
4	0.23	7
5	0.25	6
6	0.175	11
7	0.15	14
8	0.090	14
9	0.12	15
10	0.13	13
11	0.14	12
12	0.05	16

(a) Using the data in Table 3.3, plot a scattergraph to show the relationship between wave steepness and beach profile.

(b) Calculate the Spearman rank correlation coefficient and its statistical significance. Comment on your result.

It has been observed that the same beach undergoes a changing profile on an annual cycle (Figure 3.21).

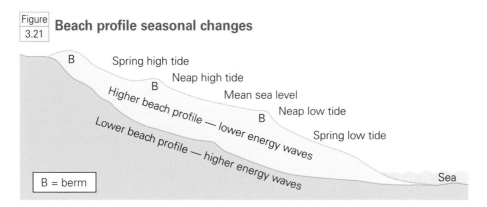

Figure 3.21 **Beach profile seasonal changes**

The more energetic waves of winter erode and transport sediment offshore, perhaps forming an offshore bar. During the summer, less energetic waves move material onshore, building the beach up into a steeper profile. This simplification is fine up to a point, but high-energy waves also occur in summer and low-energy waves occur in winter. Terms such as destructive and constructive as applied to waves are not helpful in this context. It is the amount of energy that a wave brings onto a beach that is important.

In some regions, the terms 'summer profile' and 'winter profile' are not appropriate. Coasts affected by monsoons often show clear profile changes with the wet and dry seasons.

The relationship between beach profile and wave energy is complex. Waves moving onto a steep beach tend to break directly on the beach face by surging or plunging. A significant amount of the incoming energy is reflected back from the beach, leading to such beaches being known as **reflective beaches**. Shallow-angled beaches receive waves that break and spill as the wave base is reached further out from the shore. Incoming wave energy is dissipated as the waves move across the wide beach, leading to these beaches being called **dissipative beaches**.

The role of beach material: sediment size

As with all slopes, the size or calibre of the slope material influences its shape and steepness. Steep beaches are associated with larger sediments; shallow profiles, with finer sand. The link between sediment size and the gradient lies in contrasting percolation rates. **Percolation** is the rate at which water drains through a material.

With coarse-grained sediment, larger spaces between individual particles allow water to pass through more rapidly. Therefore, as swash moves up a shingle beach, much water drains down through the beach. As a result, backwash is reduced in strength. In terms of sediment transport, there is more energy available to move particles up the beach. The return flow of water is limited, so there is little energy available to carry sediment seawards. The overall effect is to build up the beach, thereby steepening the profile.

Smaller-calibre sediment, such as sand, has less space between particles. Less water drains through the beach during swash flow, resulting in more water returning down the beach as backwash. More energy is available to transport sediment in the backwash. This lowers the gradient of the beach profile.

The role of the **water table** in a beach is important and is linked to sediment size. The water table represents the upper level of saturation beneath the beach surface. Water drains easily through coarse sediment, resulting in a water table well below the surface. This leaves an unsaturated zone into which additional water can drain. Less water flows as backwash on such beaches. Sand on the other hand, tends to be saturated almost to the surface. Little additional water can drain away, so more water flows across the surface (similar to the conditions that result in overland flow within a drainage basin). More water flowing across a beach surface means more transport of sediment. In particular, a higher ratio of backwash to swash tends to move material offshore, thus lowering the beach gradient.

Particle size affects the angle of rest — the larger the particle size, the steeper is the angle of rest. Thus, pebbles form a steeper slope than can sand; coarse sand a steeper slope than fine sand.

On beaches that store both shingle and sand, all these factors operate. There is often a marked break of slope between the steeper shingle segment and the lower-angled sand segment (Figure 3.22).

| Figure 3.22 | **Sand-and-shingle beach, Norfolk** |

Fine-calibre sand forming very gentle slope

Coarse-calibre shingle forming steeper slope

Break of slope

Michael Raw

The role of beach material: sediment shape

Sediment shape influences its movement (see page 46). By using the dimensions of the three axes, larger particles such as pebbles can be placed in one of four shape categories:

- disc — round and flat
- sphere — ball-like
- rod — long and thin
- blade — long and flat

The three-dimensional shape of a particle influences its movement. Rod- and sphere-shaped particles roll easily. A disc-shaped particle can be picked up and thrown by waves pushing against its flat surface. Blades can roll but not as well as rods and spheres; they are not thrown as effectively as discs.

Activity 5

Table 3.4 shows the distribution of pebbles of different shapes on the foreshore (inter-tidal) and storm ridge (top of beach) zones of Criccieth beach. Criccieth beach is a shingle accumulation on the northern shore of Cardigan Bay (see Figure 3.7, p. 49).

Table 3.4 Distribution of pebble shapes, Criccieth beach, north Wales

Shape	Observed foreshore	Expected foreshore	Observed storm ridge	Expected storm ridge	Observed totals
Discs	8	20	32	20	40
Blades	19	30	41	30	60
Rods	33	25	17	25	50
Spheres	40	25	10	25	50

(a) Use the chi-squared test to test the hypothesis that the distribution of observed pebble shapes between foreshore and storm ridge is no different from an expected even distribution.

(b) Suggest reasons for the result of the chi-squared test.

The role of tidal range

As well as the daily variations in low and high water, tides follow an approximate 14-day cycle of high (spring) and low (neap) tides. The difference between the high and low tides influences beach profiles because it determines beach width. Where coasts receive a combination of high wave energy and have a macro-tidal range, beaches tend to be wide. These conditions are found in the British Isles, the west coast of Canada and southern South America.

On beaches made up of shingle and pebbles, distinctive ridges or **berms** can be identified. These are usually related to the levels of high water reached in the tidal cycles (Figure 3.21). At the highest level, the berm is often a flat-topped feature that changes shape because of wave energy only when a severe storm occurs at the same time as a spring tide. Sand beaches tend not to have such marked ridges in their profiles. Sand is relatively easily re-worked by the wind, so a small ridge left at a spring high water level is unlikely to last intact until the next time the tide reaches this point.

Estuaries and deltas

Estuaries and **deltas** are locations where rivers extend into the coastal zone. They result from interaction between marine and fluvial processes and are important stores of sediment. They have a variety of shapes. However, there is a fundamental difference between them:

- Deltas are areas where sediment is accumulating out from the land into the sea.
- Estuaries are indentations in the coastline, often funnel-shaped, that are infilling with sediment (Figure 3.23).

Deltas and estuaries are important locations for human activity. Therefore, changes to the processes operating in and on them, as well as the landforms themselves, are significant. The storage of weathered and eroded material from land areas in submarine sediments has a significant influence for global climate change. Such sediments can contain significant quantities of carbon and methane. Locked away in estuary sediments they cannot contribute to the enhanced greenhouse effect.

| Figure 3.23 | **Mawddach estuary, mid-Wales, looking inland** |

Peter Stiff

Estuaries

Many estuaries can be divided into three compartments (Figure 3.24).

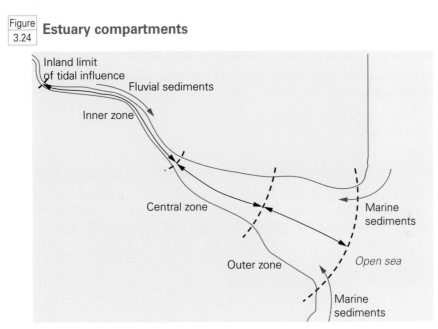

Figure 3.24 **Estuary compartments**

In terms of energy inputs, the outer (seaward) compartment receives much wave and tidal-current energy, while the inner (landward) compartment has considerable energy inputs from river currents (Figure 3.25).

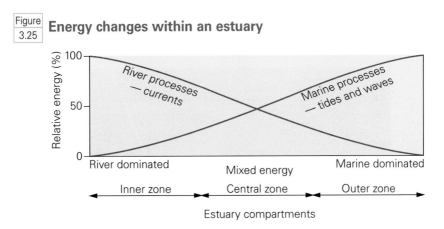

Figure 3.25 **Energy changes within an estuary**

Finer sediments are transported through these two areas into the less energetic middle compartment. It is here that most deposition occurs. Coarse material, for example sand and gravel, tends to be deposited in the inner and outer compartments.

The mixing of salt with fresh water is an important but variable process within estuaries. Mixing takes place due to diffusion and advection. Diffusion is a chemical process caused by differences in the ionic makeup of salt and fresh water. **Advection** is a physical process that mixes salt water and fresh water due to the flows of water within the estuary. In any one estuary, both processes occur at the same time, but one often dominates the other.

The degree of mixing of fresh water and salt water allows us to identify three types of estuary:
- stratified — very little mixing
- partially mixed
- well-mixed

Stratified estuaries are found in micro-tidal environments where both the tidal and river currents are not strong enough to cause the turbulence necessary to bring about physical mixing. Fresh water lies above salt water, forming two wedges: fresh water tapers seawards, salt water tapers landwards.

As river and tidal currents increase, the degree of mixing increases. The resulting patterns of fresh, brackish (a mixture of salt and fresh water) and salt water influence ecosystem development.

Although there are significant flows of water into, out of and within estuaries, the dominant process is deposition. Therefore, the descriptive term **sediment sink** is given to an estuary. The stores of sediment, for example sand and mud banks, can frequently shift location (Figure 3.26).

Figure 3.26 **Deposition in the Kent estuary, Cumbria**

Older infill and land reclamation

Fluvial sediments transported by River Kent

Salt marsh and recent sedimentation

River Kent (shifting channel)

Sand flats exposed at low tide

cleo2.lancs.ac.uk

Flood tide sediments from Morecambe Bay

Deltas

These accumulations of river-derived sediment occur at a variety of scales. A small stream entering the sea can develop a delta of a few tens of square metres. At the other extreme, there are the deltas of continental-scale rivers such as the Nile, Mekong and Mississippi, that cover hundreds of square kilometres.

Coastal deltas are the result of a combination of the following factors:

- river energy
- sediment transported by river
- wave energy
- tidal range and flows
- shelf width and gradient
- tectonics

The relative influences of these factors has led to a classification of deltas (Figure 3.27).

Figure 3.27 **Types of delta**

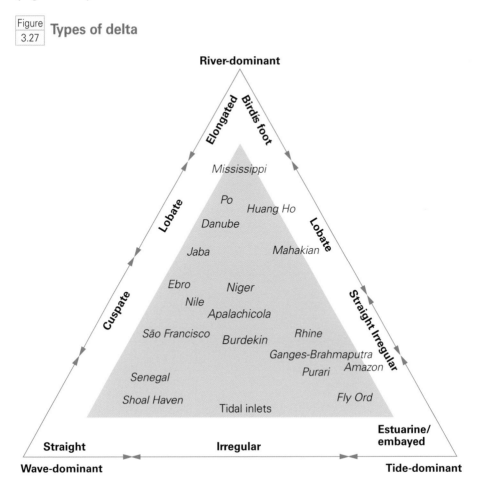

- **River-dominant** deltas are characterised by large catchments, with the river discharging into a relatively low-energy sea area. Examples are deltas of the Mississippi and the Danube.
- At locations where tidal energy is high and wave energy is moderate, sediment is deposited perpendicular to the coastline. Here, **tide-dominant deltas** form. Example are those of the Ganges–Brahmaputra and the Rhine.
- Deltas that front onto areas of open sea and receive high levels of wave energy have smooth coastlines. Strong longshore currents operate and contribute to the formation of linear accumulations of sediment parallel to the coastline. These **wave-dominant** deltas often have well-developed beaches and sand-dune systems. The São Francisco river in northeast Brazil has an example of such a delta.

However, it is important to appreciate that any one delta is unlikely to owe its shape entirely to one set of processes. Small changes in sea level alter the energy input to deltas as waves and tides extend their influence over more and more of the delta.

Most present-day deltas are relatively young landforms. Sea level has been at its current position for about the last 6000 years, so it is during this time period that deltas have developed, some involving very large volumes of sediment. In the northwest of the Indian sub-continent, two major rivers share in the production of one of the largest deltas in the world. The Ganges and Brahmaputra transport about 1.5 billion tonnes of sediment per year to the Bay of Bengal. Their delta now consists of some 1500 billion m^3 of material lying above mean tidal level and nearly 2000 billion m^3 extending out under the sea.

4 Ecosystems in coastal environments

As with any ecosystem, coastal ecosystems are **holistic** — all the components are interconnected. Change in any part of the ecosystem tends to bring about change throughout. For example, a tropical storm can cause significant coral-reef erosion, a fire can destroy dune vegetation and cliff collapse can cover an area of shore platform, smothering its community.

Plants, animals and microorganisms occupy habitats in the coastal zone that provide them with:

- mineral and organic nutrients
- rocks and sediment on which to grow
- water
- solar energy

Biotic components (living organisms) are influenced by the **abiotic** or physical environment. On the other hand, they can bring about change in their physical environment. Perhaps the most extreme situation is the building of a reef by coral to form an atoll. Plants also play a role in stabilising sediments, such as mud and sand.

Two important types of ecosystem are found in the coastal zone:

- **hydroseres** — communities found in wet or waterlogged conditions — for example, marsh
- **xeroseres** — communities found in dry conditions — for example, sand dunes

The biotic components found in coastal hydroseres and xeroseres are, to a greater or lesser degree, **halophytic**. This means that they are tolerant of relatively high levels of salt in the environment (salt is toxic to terrestrial plants and animals).

Ecological succession

Vegetation change within an ecosystem is called plant **succession**. It involves a sequence of **seres** (plant communities) in which each community is more complex than the previous one. It is suggested that a succession of vegetation types takes place over time and includes the following changes:

- increasingly favourable physical environment — for example, soil, water availability, shelter
- progressive increases in nutrient and energy flows
- increased biodiversity
- increased net primary productivity

A **climax community** is reached eventually, with plants and animals existing in a state of equilibrium with climate and soil conditions. So long as the physical environment remains stable, the climax community should persist indefinitely.

However, predictable change is rare and the dynamic nature of ecosystems is seen as important. Four sets of factors are relevant to change within an ecosystem:

- **autogenic** factors — internal to the ecosystem — for example, decaying plant material changing the soil conditions
- **allogenic** factors — external to the ecosystem — for example, climate change
- human activities — for example, pressure due to trampling
- time

Coastal ecosystems tend to be prominent where sediment can accumulate. In such locations, plants gain an initial foothold and other elements of the food chains and webs then develop.

Tidal flats, salt marshes and mangroves

Tidal flats

Tidal flats are areas of mud and sand deposits that are common in estuaries. However, they can develop in a number of locations as sediment accumulates. In general, tidal flats are covered at high tide and are exposed when the tide is out. They have a low gradient across their surface. Mud made up of clays and silts is often the main material, but closer to the low-tide level, the proportion of sand increases. At low tide in estuaries, **shoals** of mud and sand can be present as islands. Small channels of water can carve dendritic patterns across the flats at low tide (Figure 4.1).

Figure 4.1 **Laugharne estuary, south Wales**

Tom Stiff

Formation of tidal flats

Clay and silt are the smallest-calibre sediments and are transported in suspension, even when water energy is low. There is a simple argument that they are deposited in low wave-energy environments. However, this is only part of the story. The most extensive tidal flats are located in macro-tidal environments, where tidal currents have more than enough energy to keep these very small particles in suspension. The period of slack water, around low and high tide, is typically short and the settling time required for fine clay and silt sediment is long. Large volumes of water are moved into and out of an area of tidal flats with the rising and falling tides. Some estuaries also receive significant inputs of wave energy.

Clay particles are not tiny grains of larger particles such as sand. Clays are formed by the chemical processes of weathering. The particles possess negative charges that cause the individual particles to stay apart in fresh water. As clay enters an estuary, the positive sodium ions in saline water overcome the repelling negative forces of the clay particles. When individual clay particles come close, they bind together to form larger agglomerations called **flocs**. The process of **flocculation** helps to explain the accumulation of clay in salt marshes and mudflats.

As well as this electrochemical flocculation, organic flocculation may take place. At locations where there are many invertebrates, clay particles are ingested by these organisms. The invertebrates digest any organic matter in the clay and then excrete pellets that are essentially clay flocs. The organic flocs are large enough to settle out and be deposited.

Activity 1

The mean tidal range (m) and the mean rate of net accumulation of salt marsh (cm) at various locations are given in Table 4.1. The rate of net accumulation is the mean annual accumulation of salt marsh minus the local relative sea level rise.

Table 4.1 **Rates of salt marsh accumulation and tidal range**

Salt marsh	Mean tidal range (m)	Mean rate of net accumulation (cm)
1	0.9	1.0
2	1.2	2.0
3	1.3	2.5
4	1.6	0.5
5	1.8	3.0
6	2.0	3.0
7	2.1	2.0
8	2.2	1.5
9	2.8	4.5
10	3.0	6.0
11	3.0	8.0
12	3.4	5.9
13	4.0	1.5
14	4.1	4.0
15	4.4	7.0
16	3.9	15.0

(a) Using the data in Table 4.1, plot a scattergraph to show the relationship between mean tidal range and mean rate of net accumulation.
(b) Calculate the Spearman rank correlation coefficient and its statistical significance. Comment on your results.
(c) Suggest why the rate at which salt marshes accumulate might vary over time.

Salt marsh ecology

In the parts of the tidal flats closest to the high tide level, **halophytic** (salt-tolerant) plants can invade and succeed (Figure 4.2).

Figure
4.2
Salt marsh, Arnside, Cumbria

Michael Raw

Pioneer species such as marsh samphire (*Salicornia* spp.) and cord grass (*Spartina* spp.) can tolerate the harsh conditions in such habitats. Factors such as the following impose severe limits on plant colonisation:

- flooding twice daily by tide
- high levels of salt
- strong winds
- wave action

Plant stems slow water flow, which creates conditions for sediment to settle out. This occurs not only at the lowest part of the tidal cycle but also during the periods either side of low tide. Deposition takes place for longer when plants are present. The height of the marsh surface increases; rates of about $2\,\mathrm{cm\,yr^{-1}}$ are not uncommon. Conditions change with this increase in height (Figure 4.3). For example, the sediment is drier for longer and is less salty. Different plants are then able to invade, such as sea purslane (*Halimione* spp.), salt marsh grass (*Puccinellia*) and sea lavender (*Limonium* spp.). As the marsh builds up, areas that experience only short periods of flooding by salt water are colonised by plants such as sea rush (*Juncus maritimus*).

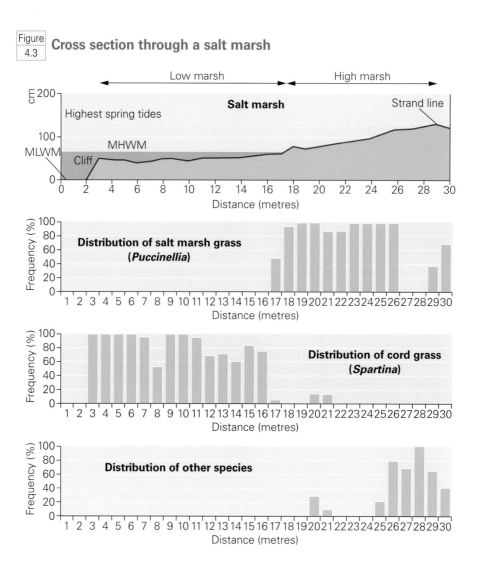

Figure 4.3 Cross section through a salt marsh

Along with plants, land-based organisms such as insects and other animals colonise the marsh. Marine species (crabs and shellfish) dominate in the lower marsh. With increasing distance from the low-water mark, land-based insects and larger animals are more common. In some locations, the marshes are used for grazing sheep.

Tidal flats and salt marshes overview

You might think that tidal flats and salt marshes seem uninviting locations for ecosystem development. However, compared with other ecosystems, they often have high rates of net primary productivity. At their most productive, these locations are only just behind the most fertile farmland and virgin tropical rainforest. The mud contains nutrients that support vast numbers of organisms, for example worms and crustacea. The mixing of salt water and fresh water in estuaries, where tidal flats and salt marsh frequently form, provides a nutritious 'soup', allowing a diverse set of food chains and webs to develop. The large flocks of wading birds that congregate at locations such as the Exe estuary and The Wash are evidence of the productivity of tidal flats and salt marshes.

Activity 2

Using either a 1:50 000 or 1:25 000 OS map, trace a stretch of coastline that includes areas of mudflat and/or salt marsh. Annotate the map so that it describes and explains the inputs and processes that led to the formation of the tidal flats and salt marsh.

Mangroves

Mangroves are a group of tree species that can tolerate relatively high levels of salt water. Where there are extensive areas of these trees, **mangrove forests** or **mangals** develop. They are restricted spatially to a zone up to about 30° either side of the equator. Mangroves often have multiple aerial roots that emerge from the trunk above the mud, anchor the tree, help with oxygen uptake and assist in the trapping of sediment. As with tidal flats in temperate latitudes, flocculation is an important process. Mangrove forests are, therefore, important stores of sediment.

Three major types of location in which mangrove forests occur are:
- river-dominated deltas (e.g. Niger delta, west Africa and Mekong delta, southeast Asia)
- tide-dominated estuaries (e.g. in Northern Territory, Australia)
- coral-reef islands (e.g. Grand Cayman, Greater Antilles)

There tends not to be a strict vegetation succession in mangrove forests. Different species colonise different types of location within the delta or estuary. A mangrove forest is a diverse ecosystem with high primary productivity. In a similar way to salt marshes at higher latitudes, mangroves act as nurseries for invertebrate species and fish. The main contrast with salt marsh is that mangrove forests have a much greater proportion of biomass above the water level.

Mangroves act as effective barriers, absorbing wave energy and, therefore, protecting the shore from erosion. However, they are prone to destruction by tropical storms. Where this occurs, or if they are removed by human activity, there is an increased hazard risk for the coastal zone. An example of this is in Bangladesh.

Benefits of salt marshes and mangroves

Together, salt marshes and mangroves offer four key benefits:
- production of habitat and fish for fisheries
- removal of impurities from water
- aesthetics
- reduction in risk from hazards

Marine dunes

Coastal sand dunes are common features of many coasts in mid-latitudes. They develop above the high-tide level and can extend as far as 10km inland. Some dune systems consist of a sequence of ridges and troughs parallel to the shoreline. Others are more complex. Ridges at right angles to the sea, or bending away from the beach, can be found. The height of the ridges varies from 1–2 m up to 30 m. Several factors need to coincide for a dune system to be formed:
- an abundant supply of sand
- a low beach gradient
- a macro-tidal range
- strong onshore winds
- an area inland of the beach where dunes can develop
- vegetation to colonise dunes

Activity 3

(a) Describe, using a labelled sketch, a beach that has a low gradient and a high tidal range.

(b) Annotate the diagram to explain the significance of these two factors in the formation of marine sand dunes.

Transport of sand is mainly by **saltation**. On land, this is an energetic process because the falling sand grain has a greater impact than it does in water. Also, sand grains are projected high above the surface — by as much as 1 m. Wind velocity increases significantly above the surface, so grains are transported readily downwind. Once sediment is moving, a small increase in wind velocity generates a large increase in sand transport. For example, a 25% increase in wind speed brings about a doubling of the sediment transport rate.

Grains can also be rolled over a flat surface and slide down slopes.

Activity 4

Study Figure 4.4, which shows vertical changes in wind speed above bare sand (a) and where vegetation is present (b).

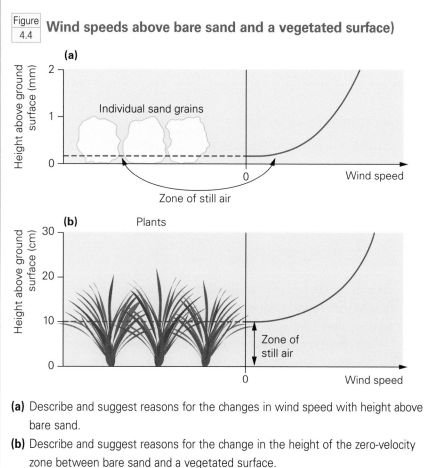

Figure 4.4 **Wind speeds above bare sand and a vegetated surface)**

(a) Describe and suggest reasons for the changes in wind speed with height above bare sand.

(b) Describe and suggest reasons for the change in the height of the zero-velocity zone between bare sand and a vegetated surface.

(c) For both bare sand and vegetated environments, discuss the importance of the zero-velocity zone in the process of saltation and the accumulation of sand.

Formation of marine dunes

Marine dunes usually begin to form above the spring high tide level. Obstacles such as accumulations of seaweed or driftwood cause wind speed to fall in their lee. Sand grains are deposited in this area of reduced wind speed. If **pioneer plant species** establish themselves here, the stems and leaves of the plants reduce wind speed through friction, so more sand accumulates. Roots also act to trap sand. Small and low **embryo dunes** develop. If sufficient sand accumulates, neighbouring embryo dunes merge to give a line of **foredunes** about 2 m high that marks the back of the beach (Figure 4.5). As these are colonised by plants, they grow both vertically (up to about 10 m) and in width, forming a substantial ridge. In cross section, they have steeper slopes facing the wind and gentler lee slopes facing inland. Wind speeds are high on the windward side and on the ridge but are reduced on the lee side. Saltating sand grains move up and over the ridge and are deposited on the lee slope. The whole dune moves gradually inland. Rates of movement as high as $7\,\text{m}\,\text{yr}^{-1}$ have been observed.

Figure 4.5 **Dune system cross section**

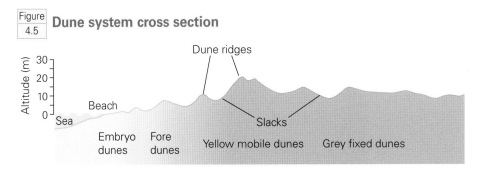

A sequence of parallel ridges can develop extending inland. Between ridges, hollows (**slacks**) are found (Figure 4.6). As soon as the wind crosses a ridge, its speed at ground level falls and then increases again towards the bottom of the lee slope. Wind speed is high enough to pick up and transport sand. This makes erosion the dominant process. Some slacks are deep enough for the water table to reach the surface. This tends to happen seasonally, for example during the winter at mid-latitudes, when rainfall input is high.

Figure 4.6 **Dune ridge and slack, Ainsdale, Lancashire**

With distance from the sea, dune height decreases because less sand is transported this far from the beach. The lines of these ridges can be broken by **blowouts** — areas of bare sand (Figure 4.7).

Figure 4.7 **Dune blowout, Sandscale, Cumbria**

Michael Raw

The formation of blowouts is often initiated by the removal of significant amounts of vegetation. The wind then removes sand by **deflation**. This is an example of positive feedback because the loss of vegetation allows sand removal, which leads to increased wind speed as friction from plants is reduced. Higher wind speeds keep sand grains mobile and make it hard for plants to establish. Therefore, the area of bare sand expands.

Causes of blowouts (deflation hollows) include:

- animal activity (e.g. rabbit burrows)
- wave activity (e.g. storm waves in combination with a particularly high tide)
- human activity (e.g. trampling of vegetation by walking, riding, driving)

Activity 5

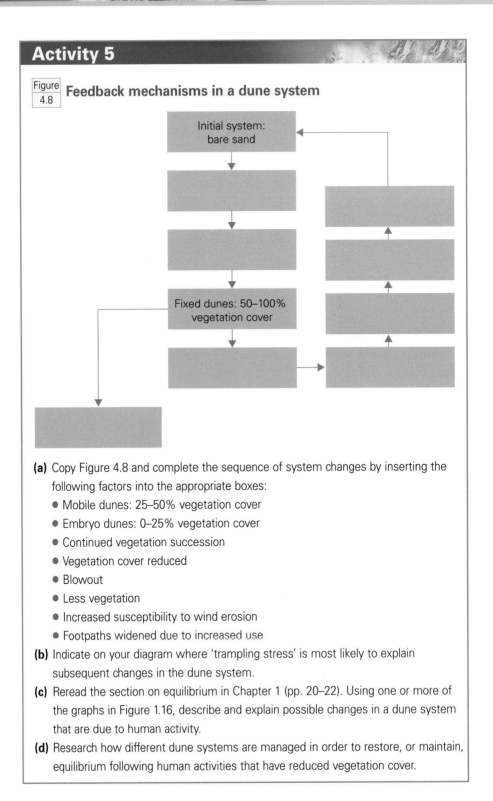

Figure 4.8 Feedback mechanisms in a dune system

Initial system: bare sand

Fixed dunes: 50–100% vegetation cover

(a) Copy Figure 4.8 and complete the sequence of system changes by inserting the following factors into the appropriate boxes:

- Mobile dunes: 25–50% vegetation cover
- Embryo dunes: 0–25% vegetation cover
- Continued vegetation succession
- Vegetation cover reduced
- Blowout
- Less vegetation
- Increased susceptibility to wind erosion
- Footpaths widened due to increased use

(b) Indicate on your diagram where 'trampling stress' is most likely to explain subsequent changes in the dune system.

(c) Reread the section on equilibrium in Chapter 1 (pp. 20–22). Using one or more of the graphs in Figure 1.16, describe and explain possible changes in a dune system that are due to human activity.

(d) Research how different dune systems are managed in order to restore, or maintain, equilibrium following human activities that have reduced vegetation cover.

Loose sand from a blowout can be redeposited in a U-shaped **parabolic** dune. Such dunes tend to be found at the furthest point inland of the dune system.

Dune systems can be looked at in terms of their development through time. The greater the distance from the sea, the older the dunes are, so there is also an interesting spatial pattern. This means that a dune system can be investigated, both temporally and spatially, by following a transect across the dunes.

Activity 6

This fieldwork activity aims to:

- measure the cross-section profile of a dune system
- compare changes in the shape of successive ridges and troughs across a dune system

Select a suitable area of dunes. Human influences should ideally be minimal, but in many locations this is difficult because of past use and current management schemes. Carry out some background research, including local history sources, to try and establish the known evolution of the dunes. You must also consult with local parties such as landowners or conservation bodies regarding access. (Make sure you complete a full risk assessment of each proposed location.)

(a) Chose a starting point for the transect on the beach, several metres from where the embryo or foredunes begin.

(b) Work along the transect running as close to 90° to the line of the dune ridge as possible. Use a compass to keep the transect on course.

(c) Measure the angle of the slope segment between each break of slope. Remember to devise a recording sheet that clearly distinguishes between positive and negative slopes. Make notes and take photographs as you walk through the dune system.

The end point of the transect is often pre-determined when a land-use, for example farming, housing or a golf course takes over.

(d) Draw the cross-section profile on graph paper. Remember to choose sensible vertical and horizontal scales. For a true representation of the dune system, the two scales would be identical. However, to see the profile clearly, it is necessary to exaggerate the vertical scale. Make sure that you do not overcompensate.

(e) Compare the changes along the transect in terms of:

- slope angle
- height of dune ridge
- depth of trough or slack
- changing relationship between ridge and trough

Annotate your cross section to highlight the key changes in shape across the dune system.

(f) Evaluate your methodology.

The ecology of marine dunes

A beach is a hostile environment for most plants. Those that manage to survive possess adaptations that allow them to overcome such factors as:

- scarcity of fresh water
- little water retention in the soil
- mobile land surface (sand)
- high levels of salt
- exposure to strong and persistent winds
- extreme diurnal surface temperatures
- low levels of nutrients

The pioneer plants that manage to colonise embryo dunes and those that dominate the other dune ridges are tough varieties with a range of adaptations. Many species are **xerophytes** — plants adapted to dry environments. In addition, those plants found closer to the sea tend also to be **halophytes** because of their tolerance to salt-rich environments (Table 4.2).

Table 4.2 **Dune plants: their locations and adaptations**

Location within dune system	Characteristic plants	Adaptations
Embryo dunes	Saltwort (*Salsola kali*)	Succulent
	Sea bindweed (*Calystegia soldanella*)	Creeps along surface; readily roots from stems in contact with surface
Foredunes	Sand couch (*Agropyron junceiforme*)	Spreads by underground rhizomes; narrow leaves; wind pollinated
	Sea spurge (*Euphorbia paralias*)	Succulent; spreads by underground rhizomes
Yellow dunes	Marram grass (*Ammophila arenaria*) — see Figure 4.9	Very long roots; spreads by underground rhizomes; narrow leaves curl round; ridged leaf surfaces
Grey dunes	Dandelion (*Taraxacum officinale*)	Long tap root; low-growing leaf rosette; wind-dispersed seed
	Red fescue (*Festuca rubra*)	Spreads by underground rhizomes; narrow leaves; wind pollinated
	Gorse (*Ulex* spp.)	Thorn covered; regenerates if burnt
Slacks	Heather (*Calluna vulgaris*)	Woody stems; regenerates if burnt
	Rush (*Juncus* spp.)	Tolerates waterlogged soil
	Alder (*Alnus* spp.)	Small tree; tolerates waterlogged soil

Local conditions within dune systems vary, so the distribution of plants in Table 4.2 is only an approximation. It is also a simplification, as there are many other species found within a dune system. Such plant communities are known as **psammoseres**. They are a specialist type of xerosere. Insects and other animals are also part of this ecosystem. Dune slacks are important habitats for rare species, for example natterjack toads and some orchids.

Figure 4.9 **Marram grass, Dawlish Warren, Devon**

Peter Stiff

Activity 7

This fieldwork activity aims to:
- measure changes in vegetation across a dune system
- measure changes in soil conditions across a dune system

(a) Refer to Activity 6, investigating changes in the profile of a dune system. The same points concerning choice of dunes, background and risk assessment apply. This activity, or selected elements of it, could be carried out at the same time as the previous activity.

(b) Establish a transect line across the dunes.

(c) Begin with a location on the upper beach and collect data at each survey point along the transect. These points could be in the middle of the slope segments used for the previous activity. Alternatively, they could be at points representing the main parts of the dune system, for example at each ridge and each trough. It is important to record the distance from the start of the transect.

The data to collect are:
- soil pH
- sample of soil for subsequent analysis of moisture content and organic content
- colour of soil
- vegetation type and degree of cover

(d) Plot the data for pH, moisture content and organic content as scattergraphs, with distance along the *x*-axis.

(e) Plot the data for vegetation change using kite diagrams along the line of transect.

(f) Using a correlation coefficient, analyse the statistical significance of the relationship between distance from the sea and the three variables.

(g) Comment on the results and their significance for changes in vegetation.

(h) Evaluate your methodology.

5 Sea-level change

Sea level is continuously changing — for example, with each tide. However, longer-term changes have significant implications for how the coastal system operates. A change in sea level alters energy inputs and outputs and is, therefore, important for the development of landforms. There is no length of coast that has not undergone considerable variation in sea level in the past 2 million years. In some locations, this variation has been as much as 120 m.

Some coastlines have had a sea level at or close to its current height for longer than others. For example, the sea has been around its present-day level for some 6000 years in Australia, whereas in Florida the sea level has been where it is today for only the past 1000 years. This is explained in part by the different relative movements of land and sea. Also, the surface of the oceans is higher in the tropics than in higher latitudes because the water is warmer and so is of a lower density, and, therefore, takes up a greater volume.

Causes of sea-level change

Changes in absolute sea water levels are called **eustatic** changes. They are global because all the oceans and seas are interconnected. Water is present in a variety of forms on Earth (e.g. rivers and lakes, ice, atmospheric water vapour). However, the total volume of global water — liquid water, ice and water vapour — is constant. Changes in where and how water is stored cause eustatic change. Of all the water stores, land-based ice is the most significant to eustatic change. This is because of the relative volume of water contained in ice sheets compared with other stores of water. For example, if all the land ice in the world melted, sea level would rise by some 90 m. However, if all the atmospheric water were transferred to the oceans, sea level would rise by just 36 mm. Locally, atmospheric water can make a difference in the short term. For example, the Bay of Bengal can rise by up to a metre during the monsoon season when vast volumes of water discharged from rivers, for example the Ganges, enter the sea. This additional water then disperses throughout the wider oceans and the bay returns to its long-term level.

Changes in the absolute level of the land are called **isostatic** changes. They are localised and are the result of tectonic influences or a redistribution of weight. The Earth's crust is capable of being depressed into the semi-molten upper mantle when enormous weight is added at the surface. This happens when ice accumulates during a glacial or when large quantities of sediment are deposited in a river basin such as the Mississippi. Indeed, the Mississippi delta has sunk by some 165m in the past 10000 years.

Smaller-scale sinking can result from human activity. Abstraction of water, oil and gas from underlying rocks can cause the land to subside. In the Tokyo region, a reduction in ground level of about 4.5m has occurred due to the abstraction of groundwater for domestic and industrial use.

The deposition of sediments in ocean basins also depresses oceanic crust. Hand-in-hand with this is the reduction in volume of the basin as sediments build up. Add to this the possibility of sediment removal from the oceans, either through uplift out of the water or by subduction, and you have some idea of how complicated sea-level change is.

Tectonic activity can cause land to rise or fall. Movement along a fault can lift land by several metres in a sudden event. This has occurred at locations around the active plate boundaries that mark the Pacific coastline. Along both Alaskan and Californian shores, lengths of coast have been elevated by tectonic forces. On a larger scale, the collision of India and Eurasia formed the Himalayas and the Tibetan plateau. The continental crust in this region was thickened and the continental area reduced. With more volume to fill, it has been estimated that the oceans fell by about 18m.

Ocean basins change shape over geological time. In places, ocean floors sink, increasing the capacity of the ocean to store water. Sea-floor spreading widens oceans such as the Atlantic, increasing their capacity. The growth of mid-ocean ridge systems has a significant influence on the volume of ocean basins. Present-day ridge systems occupy a volume equivalent to about 12% of the total volume of the oceans. As the lengths of ridges and their rates of spread vary, so does sea level.

Exploration by energy companies has yielded much geological information that has increased our knowledge of sea-level changes. Details are patchy, but the overall trend is well established (Figure 5.1).

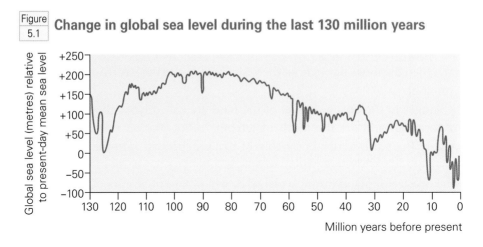

Figure 5.1 Change in global sea level during the last 130 million years

Global sea level (metres) relative to present-day mean sea level

Million years before present

Relative sea level

In many cases, it is not possible to be certain about the precise cause of an observed change in sea level. It is **relative sea-level change** that really matters, i.e. the balance between sea level and land level. A positive sea-level change, caused by either a rise in sea level or a fall in land level, gives rise to **transgressive** conditions. These lead to the drowning of coastal areas and/or the onshore migration of some types of landform, for example beaches. **Regressive** conditions result from a negative change in relative sea level. Emergent features result as the coastline builds out from its previous position.

Changes associated with glaciation

The most recent geological period, the **Quaternary**, began about 2 million years ago. The **Pleistocene Epoch** made up most of the Quaternary, and ended about 10 000 years ago. The last 10 000 years are known as the **Flandrian** or **Holocene Epoch**. During this period, the growth and decay of ice sheets and glaciers had significant effects on both absolute and relative sea levels.

The last major ice advance peaked at around 18 000 years before present when vast ice sheets extended hundreds of kilometres equatorwards from their current distribution. A glacial or 'ice age' is made up of multiple advances and retreats of ice during which sea levels rise and fall.

The main mechanism for sea-level change during glacial periods is the transfer of water from its store in the oceans to its store as ice on land during glacial periods. During inter-glacials, when ice sheets melt, the process is reversed and sea level rises. This process is called **glacio-eustasy**. There is agreement about the general pattern of sea level change, but the details are disputed (Figure 5.2).

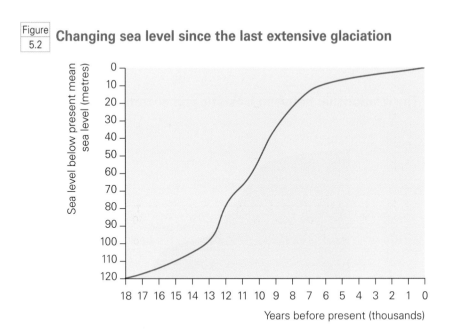

Figure 5.2 **Changing sea level since the last extensive glaciation**

The addition to, or melting of, floating ice sheets and sea ice in the polar regions has no effect on the eustatic sea level. The weight of this ice is already supported by the water and so contributes to sea level.

As ice builds up on land, crustal depression occurs; when the ice melts, weight is removed from the land and it adjusts upwards. This process is called **glacio-isostasy**.

Coasts that have experienced both types of movement (e.g. many of the coastlines of northwest Europe and northeast America) have particularly complicated patterns of sea level change. In general, glacio-eustasy and glacio-isostasy occur at different rates. Glacio-eustacy is relatively rapid in geological terms, whereas time lags involved with glacio-isostasy are long. In the Baltic region, where ice sheets were several kilometres thick, uplift of 300 m has occurred over the past 10 000 years. Isostatic adjustment is continuing at rates measured up to 10 mm yr^{-1}. Where the ice was thinner than in the Baltic region and depressed the crust to a lesser extent, for example in northern Britain, rates of uplift are lower — up to 2 mm yr^{-1}.

As positive or negative adjustments occur, energy inputs to the coastal zone change. For example, as sea level falls during a glacial period, the inter-tidal range falls lower and lower, leaving parts of the coastline no longer under the influence of marine processes. Sub-aerial processes begin to dominate. However, as an inter-glacial proceeds and sea level rises, marine processes once again act on former shorelines.

Activity 1

Study Figure 5.3.

Figure 5.3 **The relationship between isostatic and eustatic change**

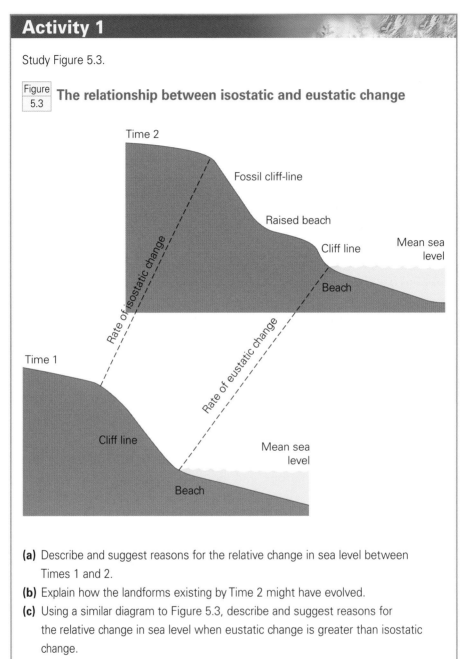

(a) Describe and suggest reasons for the relative change in sea level between Times 1 and 2.

(b) Explain how the landforms existing by Time 2 might have evolved.

(c) Using a similar diagram to Figure 5.3, describe and suggest reasons for the relative change in sea level when eustatic change is greater than isostatic change.

(d) Explain possible effects of the situation in (c) on
- a cliffed coastline
- a sand dune system
- an estuary with extensive salt marshes

Changes associated with global warming

Global warming, resulting from an enhanced greenhouse effect, is currently responsible for sea-level change. If the average atmospheric temperature rises sufficiently, ice (water stores) on the land, for example the Antarctic and Greenland ice caps, will melt and flow into the sea, causing the sea level to rise. An increase in global temperature also brings about **thermal expansion** of the oceans. The density of seawater decreases with increasing temperature, which leads to an increase in water volume. Estimates for this effect on sea level vary from 2–7 cm.

There is debate on the likely impact of ice melt on sea level. A range of estimates has been produced by researchers, such as the **Intergovernmental Panel on Climate Change (IPCC)**. Observed rises in sea level from 1890 to 1990 ranged from 10 cm up to 25 cm.

Models have been developed to predict future rises in sea level. Overall, a rise in the order of 44 cm by the year 2100 is suggested. However, the range of possible outcomes varies from 11 cm to 77 cm. Global warming over the same period is estimated to be in the range 1.5–4.5°C.

Whatever the actual changes, rising sea level is set to have significant impacts on coastal zones around the world, as the energy balance in these areas is altered. For instance, the IPCC predict that some areas will experience more frequent and intense stormy weather conditions. Therefore, more energy will be brought into the coastal zone, which will increase rates of transport and erosion.

Coastal landforms associated with rising sea levels

As sea level rises, the zone of active marine processes also rises and the coastal zone is partially submerged. The result is distinctive landforms such as **rias** and **fjords**.

Estuaries

Most estuaries around the world are drowned river valleys. These estuaries have formed in the relatively short time since sea level stabilised and a degree of equilibrium was established at the end of the last ice age (about 6000 years).

In regions where there was little or no ice on the land, the relative rise in sea level has been significant because isostatic changes due to ice loading do not occur. The term **ria** is applied particularly to drowned river valleys where the river has cut deeply into the underlying rocks. The estuary is, therefore, steep-sided (Figure 5.4).

Figure 5.4 **Part of Milford Haven ria, south Wales**

In cross section, the ria is V-shaped and in plan it has a dendritic outline (Figure 5.5). The Atlantic coastline of western Europe, for example the Iberian Peninsula and south western England and Wales, have many examples of rias.

Figure 5.5 **The Helford ria, Cornwall**

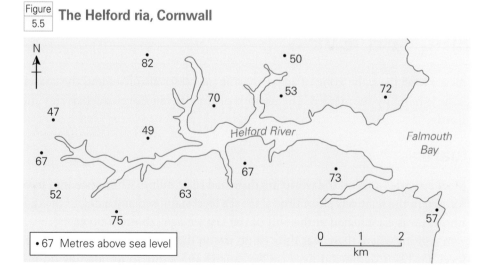

Rias are found all around the world. The coastlines of New South Wales around Sydney, and of southern Chile, possess good examples.

Where a river flows through a region of lower relief, a different type of drowned estuary forms. These are wide and shallow, and tend not to have a dendritic plan, having relatively straight, parallel banks. In cross section, they are broader and are not V-shaped. The estuaries of the east coast of England (e.g. the Thames estuary) are examples, as is the Potomac estuary in northeast USA.

In regions where ice cover was extensive, coastal valleys were altered by ice action. Valleys were straightened, deepened and left with a U-shaped cross section, creating a glacial trough. The postglacial drowning of the lower stretches of these troughs has created **fjords**. Their sides are characteristically steep (see Figure 5.6), and can reach several hundred metres above current sea level. They are also deep — 300–400 m is not uncommon (Figure 5.7). The Sogne fjord in Norway is 1000 m deep; some fjords in Antarctica reach 2000 m depth.

Figure 5.6 **Entrance to Milford Sound fjord, New Zealand**

Jane Buekett

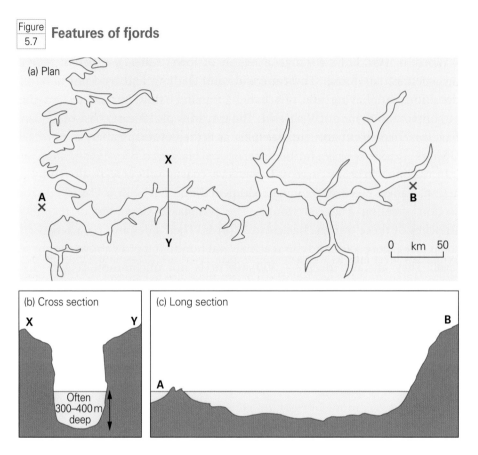

Figure
5.7 **Features of fjords**

Some fjords have pronounced sills at their mouths. The origin of this feature varies. Some sills represent glacial deposition; others are made of solid rock. They may develop where the intensity of glacial erosion is reduced as the glacier fans out onto the former shore area. Fjords can extend up to 150 km inland. Along their length, spectacular waterfalls cascade down their sides where rivers flow out of valleys perched high above sea level. These are tributary valleys left hanging along the top of the main trough once the ice melted. Fjords are common landforms along mountainous coastlines above latitudes of 45°, north and south. Classic examples include western Canada, southern Chile, south-western New Zealand and Iceland.

Drowned concordant coastlines

Where the trend of the relief of the land is parallel to the coastline, a substantial rise in sea level leads to inundation of the low-lying land. The higher land is left as islands stretching along the coast. The term **Dalmatian** style coastline (named after the Dalmatia region in Croatia) applies to this type of coastal plan.

Submerged coastlines

Around the world, there is a variety of coastal landforms lying below present-day sea level. For example, off the coastline of the Bay of Biscay, beach and sand dune systems exist at depths of between 100 m and 200 m. Around the coastline of southwest England, cliffs lie in 40–60 m of water. Former shore platforms have been found off northern Australia at depths of 200 m.

Submerged forests are further evidence of changing sea level. They are found just a few metres below present-day sea level. When low spring tides occur, their stumps can emerge briefly above the waves.

Beaches

When sea level falls as the volume of land-based ice grows, large areas of 'new' land emerge from the sea. Sediment accumulates on the surface. As the sea level rises when the ice melts, the sediment is affected by marine processes. Wave action and the rising sea level combine to push the accumulations of rounded sediments onshore. In some places, these sediments beach on a former cliff-line and build up there. Elsewhere, bars of sediment form tombolos, and barrier beaches form. The accumulation of sediment at Chesil Beach along the Dorset coast is thought to have been rolled landwards during the Flandrian transgression. Indeed, a number of the accumulations of shingle around British coasts are thought to owe more to postglacial sea level change than to processes acting today.

Coastal landforms associated with falling sea levels

As the inter-tidal zone falls lower, coastal landforms such as cliffs, shore platforms and beaches are left literally 'high and dry' — beyond the reach of marine processes. Sub-aerial processes become more influential. Such coasts are often referred to as **fossil** coasts.

Isostatic recovery following deglaciation can be responsible; higher sea levels in the past may have had a similar effect. Dating and sequencing relative falls and rises of sea level is not a straightforward task.

Fossil cliff-lines

Former marine cliffs mark the line of the coast when sea level was higher than it is today. They are distinguishable when made up of rocks that resist sub-aerial processes. The igneous and metamorphic rocks of northwest Scotland help to make former cliff-lines stand out, for example along the coastal zone of Arran. In some locations, wave-cut notches, marine caves, arches and stacks can be identified.

Raised beaches

Between a present-day cliff-line and a fossil cliff-line, deposits of sediment of marine origin can be found (Figure 5.8).

| Figure 5.8 | **Emergent coastline in northwest Scotland** |

Peter G. Knight

— Fossil cliff-line

— Raised beach

Recently abandoned cliff-line due to ongoing isostatic recovery

Raised beaches can be covered by deposits, such as **head** — a mass of clay and angular stones produced under periglacial conditions during Pleistocene times. The former beach is occasionally the surface material. Soil-forming processes have operated on this material since the end of the last glacial period and sandy soils have often formed. In Cornwall, such locations have been used to raise early crops. This is possible because, in sandy soils, the soil temperature rises relatively quickly.

Activity 2

Study Figure 5.8.
(a) Draw a sketch of this stretch of coastline.
(b) Annotate your sketch to describe and explain the present-day processes acting on both the fossil and current shorelines.

6 Human activities and the environment in the coastal zone

The coastal zone lies between the sea and land, attracting a range of human activities, but restricting others. Tidal changes mean that inter-tidal locations tend not to suit permanent land uses, such as settlement, industry and agriculture. However, landward of the high-tide level, land-uses are increasingly attracted to the coast. In the inter-tidal and offshore zones, a number of resources are exploited.

The growth in population

Coastal areas have long been important for the distribution of population. Some of the earliest archaeological evidence for human habitation has been discovered at coastal locations. Over the centuries, the association between proximity to the sea and population density has become stronger. In almost all countries with a coastline, coastal populations are growing faster than the national average.

About half of the world's 6.6 billion people live within 200 km of the coast, in an area covering some 10% of the Earth's land surface. Of the inhabited continents, only in Africa do more people live in the interior than in the coastal zone. In Asian countries, some 1.5 billion people live within 100 km of the sea. Over half the Chinese population inhabits the coastal provinces, with densities along much of the coast averaging 600 persons per square kilometre. By 2025, about 75% of the residents of the USA (excluding Alaska and Hawaii), are expected to live in coastal areas, which together account for about 17% of the land area.

There are, however, significant latitudinal variations in the attractiveness of coastlines for human settlement. High-latitude coastlines, for example in northern Russia and southern Chile, are sparsely populated. In general, coastlines in the low and mid-latitudes are more densely populated. Fertile, low-altitude coastal plains, deltas and estuaries are the most densely populated coasts. The Nile and the Chang Jiang (Yangtze) deltas and the River Plate estuary have attracted millions of people.

In spite of this trend, on a local scale, some stretches of coast have relatively few people living along them. A rocky coastline with steep cliffs and limited access to the sea offers few opportunities for settlement and economic activity. For example, the stretch of Dorset coastline between Weymouth and Swanage has a comparatively low population density compared with the south coast of England as a whole.

Activity 1

Table 6.1 **Population change in the coastal states of the USA**

State	Population, 1980 (millions)	Population, 2005 (millions)	Population density 2005 (millions km^{-2})
Alabama	3.9	4.6	34
California	23.7	36.0	84
Connecticut	3.1	3.5	271
Delaware	0.6	0.8	155
Florida	9.7	17.8	114
Georgia	5.5	9.0	54
Louisiana	4.2	4.5	40
Maine	1.1	1.3	16
Maryland	4.2	5.6	209
Massachusetts	5.7	6.4	313
Mississippi	2.5	2.9	24
New Hampshire	0.9	1.3	53
New Jersey	7.4	8.7	438
New York	17.6	19.2	155
North Carolina	5.9	8.7	64
Oregon	2.6	3.6	14
Rhode Island	0.9	1.1	387
South Carolina	3.1	4.2	51
Texas	14.2	22.8	31
Virginia	5.3	7.5	69
Washington	4.1	6.3	34
USA	**226.5**	**300**	**31**

Activity 1 (continued)

(a) For each of the states named in Table 6.1, calculate the percentage increase in population between 1980 and 2005.

(b) On an outline map of the USA that shows the coastal state boundaries, use an appropriate graphical technique to show the percentage increase in population between 1980 and 2005.

(c) On an outline map of the USA that shows the coastal state boundaries, construct a choropleth map to represent the variations in population density among coastal states.

(d) Comment on the patterns shown by the maps.

Urban coastal developments

The demand for both water and land space has led to significant human intervention in the natural coastal system.

Such is the attraction of coastal locations globally that major metropolitan regions have developed there. Of the urban agglomerations projected to be over 5 million in 2015, most occupy sites on, or very close to, the coast (Figure 6.1).

Figure 6.1 **The world's largest urban agglomerations, 1950–2015**

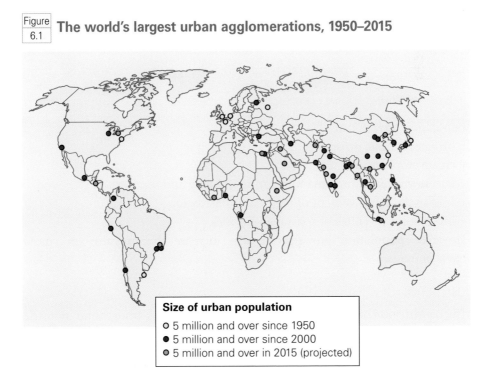

Size of urban population

○ 5 million and over since 1950
● 5 million and over since 2000
◉ 5 million and over in 2015 (projected)

Fourteen of the USA's 20 largest conurbations are coastal. Of China's 456 officially designated municipal cities, 305 are in coastal locations. In Latin America and the Caribbean, where 75% of the population are urban, 57 out of 77 major cities have coastal sites.

Most coastal cities have developed in association with port activity. As the import and export of goods grows, various feedback loops operate that encourage multiplier effects and, therefore, more growth (Figure 6.2).

Figure 6.2 **Growth of port activity**

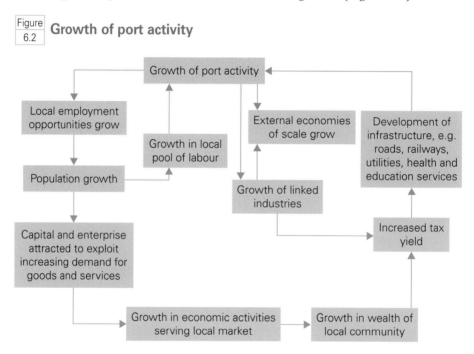

Population growth can:
- bring economic prosperity
- create employment
- encourage agriculture, manufacturing industry and services
- lead to the improvement of regional infrastructure
- increase tax revenues of coastal authorities.

However, population growth can also increase the pressure on coastal environments.

Infrastructure development

The construction of facilities such as ports, roads, railways, bridges and industrial plant and general urban development go hand in hand with the growth in population in the coastal zone (Figure 6.3).

Figure 6.3 **Port and urban development, Valparaiso, Chile**

Michael Raw

River ports, for example London and Rotterdam, developed inland from the open sea along the tidal stretches of estuaries (see Figure 6.4).

Sea ports grew along stretches of open coast. An example is the port of Los Angeles at Long Beach, California.

As soon as rudimentary wharfs and bridges are constructed, the estuary system is altered. By the seventeenth century, much of the tidal stretch of the lower River Thames was lined with a man-made shoreline of wooden walls. The original London Bridge, with its large number of stone and brick piers, reduced the flow of water and, therefore, affected estuary processes such as the transport and deposition of sediment.

By the mid-eighteenth century, docks landward of lock gates were being constructed in ports. This avoided the effects of tides on moored vessels and allowed walls to be built to keep cargoes secure in warehouses. Large areas of floodplain were taken up with dock premises and associated industry, railway links, road links and housing. As the Port of London expanded eastwards, areas

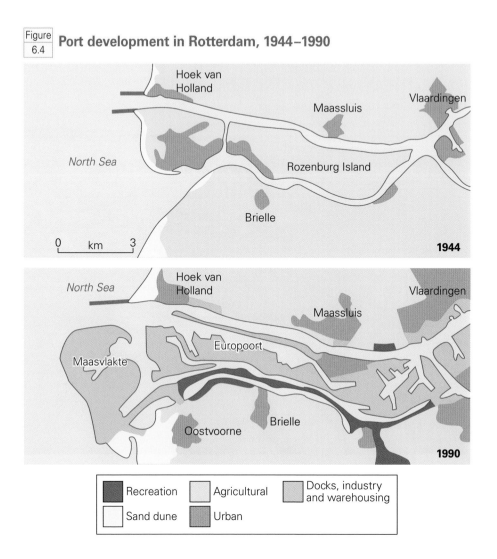

Figure 6.4 | Port development in Rotterdam, 1944–1990

either side of the Thames, for example marshland on the Isle of Dogs, were built over. Not only was the hydrological cycle affected, but fluvial and coastal processes were also altered. The flow of water was constrained between artificial banks; navigation channels were maintained by dredging excess sediment and dumping it beyond the coastal zone. Meanwhile, jetties and breakwaters modified longshore sediment drift.

Demand for land

Increasing human activity along coastlines leads to greater demand for land, with estuaries and lowland coasts placed under particular pressure. In some coastal locations, little undeveloped land remains. Therefore, inter-tidal areas are seen as offering potential, for example the area in and around Southampton. Tidal flats, sand dunes and salt marsh can be engineered to provide dry land for development. Embankments are constructed to stop the sea from entering the area. This allows drainage and the consolidation of exposed sediment for building (Figure 6.5).

| Figure 6.5 | **Wetland reclamation in and around Southampton** |

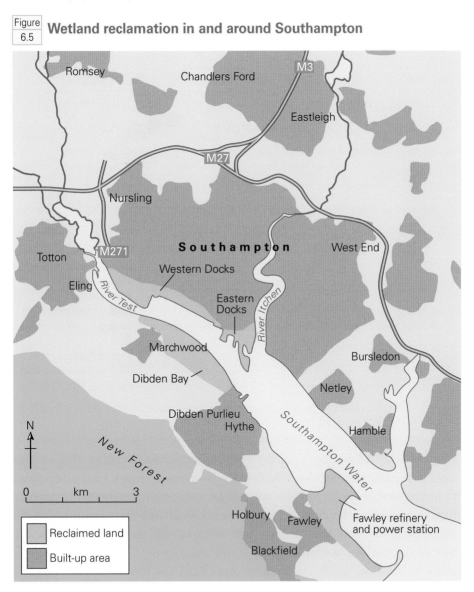

Mangrove swamps in the tropics are used for various types of food production. In Indonesia, brackish water ponds are dug to create fishponds, known locally as *tambak*. Mangrove and marshland in Hong Kong has been used for fish and shrimp production, although, more recently, rice cultivation and housing have dominated. In Singapore, shortage of space caused by the massive population and economic growth of the twentieth century has led to large-scale reclamation projects (Figure 6.6). These projects, which are on-going, have created some 50km² of additional space.

| Figure 6.6 | **Singapore: urban pressure and land reclamation** |

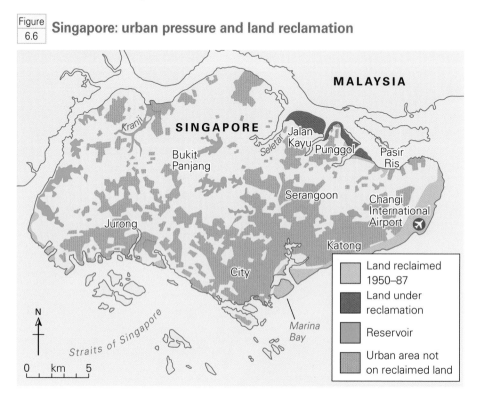

Land reclamation in the Netherlands

Some 10 000 years ago, the country now known as the Netherlands was an area of marshland, tidal inlets, tidal flats, salt marsh and sand dunes. Peat accumulated in some locations, which became less prone to flooding. Early settlers built raised mounds called 'terps' on areas of peat that had accumulated above the flood level. Under Roman influence, causeways, canals and harbours were built. The first dykes (embankments) were recorded in the tenth century. The following centuries were times of give and take between land reclamation and water inundation, when, on balance, more land was lost back to the sea than was reclaimed.

The twentieth century saw the large-scale application of technology. Schemes such as the partial draining of the Zuider Zee to create the enclosed Ijsselmeer and its dry polders (Figure 6.7), and the Delta project in the southwest, used state of the art technology (Figure 6.8). Such developments have, for the time being, given the Netherlands a degree of protection against flooding and the largest land area in its history (Figure 6.9).

Figure 6.7 **Polder landscape near Lelystad, the Netherlands**

Figure 6.8 **Dam gates are used as part of the Delta project at the mouth of the River Rhine**

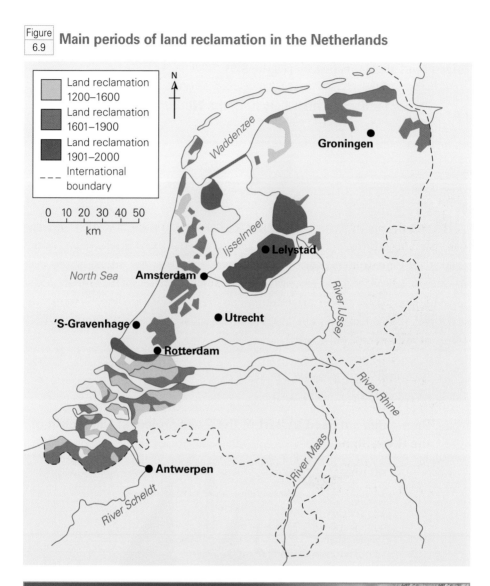

Figure 6.9 | Main periods of land reclamation in the Netherlands

Legend:
- Land reclamation 1200–1600
- Land reclamation 1601–1900
- Land reclamation 1901–2000
- – – – International boundary

0 10 20 30 40 50 km

Map labels: Waddenzee, Groningen, IJsselmeer, Lelystad, North Sea, Amsterdam, 'S-Gravenhage, Utrecht, Rotterdam, River IJssel, River Rhine, River Maas, Antwerpen, River Scheldt

Activity 2

Table 6.2 | Land reclamation in the Netherlands

Time period (yr)	Area reclaimed (km²)
1200–1600	1835
1601–1900	2790
1901–2000	1900

Activity 2 (continued)

(a) Using an appropriate graphical technique, present the data given in Table 6.2.

(b) Research possible reasons for the changes in reclamation activity in the three time periods. Note the different lengths of the time periods.

(c) Research the origins, implementation and consequences of the Zuider Zee project.

The use of resources

With so many people living within the coastal zone, resources such as **minerals** (oil, gas and sediments), **biological** resources (fish) and **space** (sand dunes, tidal flats) are under pressure. Various frameworks for managing coastal resources have been developed. Demarcation of the spatial limits to an individual country's territorial rights has been a difficult issue. While not universal, most countries use a series of zones parallel to the coast to define levels of authority (Figure 6.10).

Figure 6.10   **Territorial rights in the coastal zone**

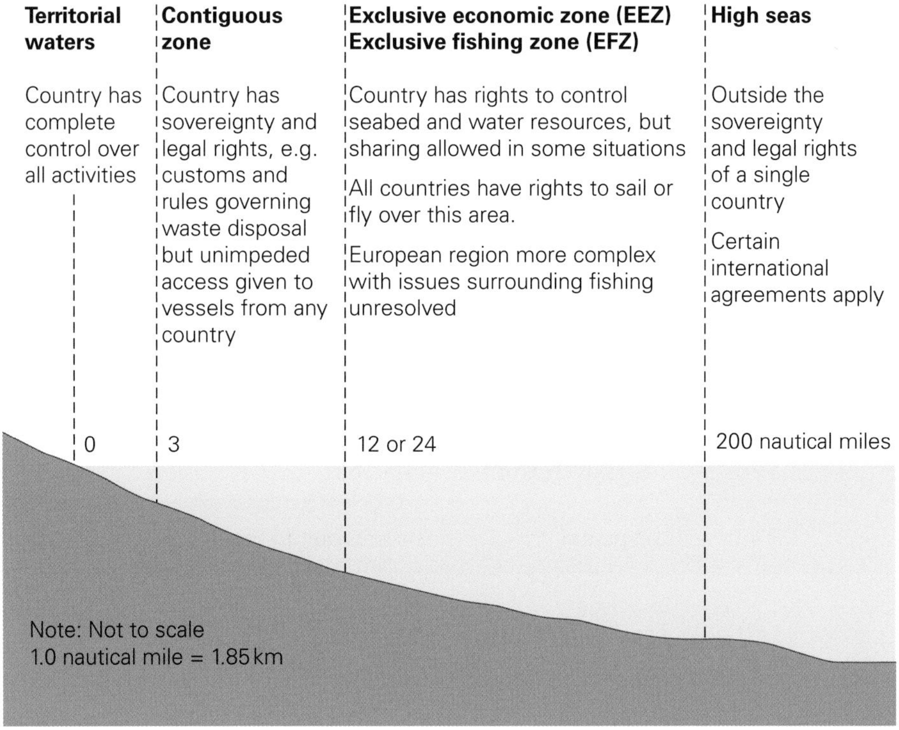

Territorial waters	Contiguous zone	Exclusive economic zone (EEZ) / Exclusive fishing zone (EFZ)	High seas
Country has complete control over all activities	Country has sovereignty and legal rights, e.g. customs and rules governing waste disposal but unimpeded access given to vessels from any country	Country has rights to control seabed and water resources, but sharing allowed in some situations. All countries have rights to sail or fly over this area. European region more complex with issues surrounding fishing unresolved	Outside the sovereignty and legal rights of a single country. Certain international agreements apply
0	3	12 or 24	200 nautical miles

Note: Not to scale
1.0 nautical mile = 1.85 km

The situation in Europe is complicated because of the Common Fisheries Policy of the EU. Under this, exclusive fishing zones (EFZs) are created. Within its EFZ, a country controls wildstock fish resources on a quota–share basis. Exact boundaries are disputed, as is their operation.

Mineral extraction: sand and gravel

Digging or dredging for aggregate, sand and gravel takes place at varying scales, from small-scale abstraction by hand digging to large-scale commercial dredging operations. This activity takes place normally only in areas licensed by governments. Several countries, including France, Japan and the Netherlands, restrict dredging landward of the 20m isobath (line of equal water depth).

Sand is used in the manufacture of glass and, along with gravel, for construction. Sand and gravel are also used in coastal management, for example in beach nourishment. Sand of different calibres is required for different purposes. The quality of sand, in terms of its cleanliness, is also important. Some sand has much fine silt mixed with it, which reduces its value because it has to be washed before use.

Dredge mining is carried out from a boat fitted either with a loop of buckets that scoop up sediment or with a powerful hydraulic system that sucks up sediment from the seabed through a metal tube (Figure 6.11). Dredge mining is likely to have the greatest impact because it removes large quantities of material in relatively short periods of time.

| Figure 6.11 | **Some of the effects of dredge mining** |

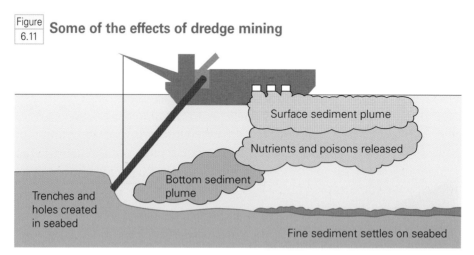

Surface sediment plume

Nutrients and poisons released

Bottom sediment plume

Trenches and holes created in seabed

Fine sediment settles on seabed

Impact of sand and gravel extraction on marine ecosystems

Disturbance of the seabed adversely affects the habitats of fish, invertebrates and algae, as well as physically removing bed-dwelling fauna. Food webs and chains are interrupted, which in turn impacts on birds and mammals. Crucially, some dredged areas are the spawning grounds for fish. Dredging can, therefore, lead to a decline in fish stocks.

In the course of dredging, small-calibre material that is not required is released back into the water. This fine sediment settles back on the seabed and can suffocate filter-feeders, for example mussels. Fine material can also fill crevices where shellfish, for example lobsters, and some species of fish live. Fish species such as herring and sand eels only spawn where the seabed comprises coarse-grained material, so the deposition of fine material reduces breeding potential.

Poisons, such as heavy metals and hydrogen sulphide, may be released from seabed sediment when dredging occurs. Plankton blooms can develop in response to the release of nutrients from the 'sea soil'. Light levels and visibility are thus reduced, affecting the ecosystem.

Impact of sand and gravel extraction on marine environments

Disturbance, similar to that produced by dredging, happens naturally during intense storms. Due to increased wave energy, the wave base increases into deeper water, and at lesser depths more energy reaches the seabed. However, the churning of the seabed lasts only a few days and the system returns to equilibrium. Dredging, however, continues for much of the year, and the resulting disturbance to the system causes permanent damage to marine environments and ecosystems.

Depending on local circumstances, extraction of sediment can change the pattern of wave energy reaching the coast. As wave crests pass over an area where sediment has been removed, some refraction occurs. Sediment removal deepens and smooths the seabed, which allows waves to pass over with less frictional interference. Therefore, wave energy decreases in the lee of the site and increases either side of this.

The removal of sand and gravel can also disturb local sediment budgets. Beach and dune systems can be diminished and, therefore, offer less protection from wave energy to the land behind them. The management of sand mining seeks to minimise effects on coastal systems. In general, the deeper the water in which the extraction occurs, the less likely is disturbance to beach and dune systems.

Mining sea sand in New Zealand

The growing demand for sand has been met by a number of offshore sources. In some instances, mining has led to positive feedback and has caused environmental problems. For example, on the North Island of New Zealand, mining off Whangateau harbour was blamed for the Omaha spit changing shape and eroding. Equilibrium was only restored by groyne construction along the spit.

As a result of the high demand for sand in the Auckland metro region (a third of New Zealanders live here) and sand's high bulk and low value, sourcing occurs close to the area. The nearshore zone just off Pakiri Beach some 100 km north of Auckland has extensive sand deposits covering an area of 500 km² (Figure 6.12). The sand is a legacy of eustatic changes towards the end of the ice age.

Figure 6.12 **Map showing location of Pakiri Beach, North Island, New Zealand**

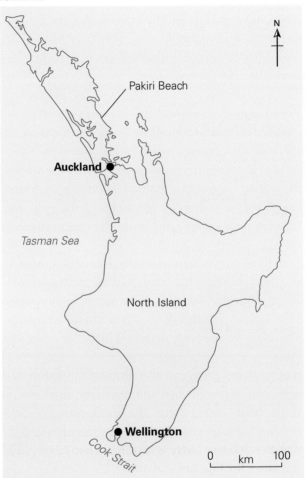

The rise in sea level resulted in the transportation of this particularly pure sediment, known as Holocene sand, to its present location. It overlies older sediment — Pleistocene sand — that is tainted with iron. Holocene sand extends some 2 km out from the shore and, although only 20–30 cm thick, is part of the dune–beach sediment system (Figure 6.13).

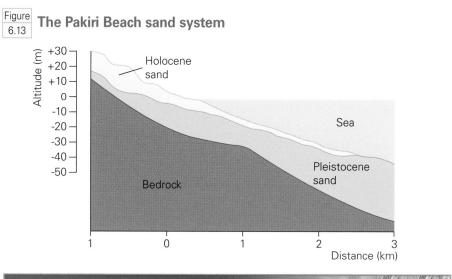

Figure 6.13 The Pakiri Beach sand system

Activity 3

Active mining of sand occurs at Pakiri Beach. It is interesting to consider to what extent the reserves of sand can be considered renewable.

(a) The river catchments supplying sediment cover 20000 hectares (200km²). Assume a rate of erosion of 1m³ per hectare per year. Calculate the volume of material transported into the sea by rivers.

(b) Not all the material transported by rivers is sand. Assume sand makes up 30% of the volume. Calculate the volume of sand brought into the coastal system around Pakiri Beach per year.

(c) Some 100000m³ of sand are mined annually. Calculate the net amount of sand extracted each year from the coastal system.

(d) The estimated volume of Holocene sand on the seabed is in the range of 82–142 million m³. Calculate the number of years the sand reserve will last for each end of the estimated range. Assess the degree to which the sand reserves in this region are sustainable.

The aim at Pakiri Beach is to restrict mining landwards of the 25m isobath and, therefore, to not affect the beaches and dunes. The dunes are under more direct threat from forestry and housing and the nearshore waters are polluted occasionally by discharges of muddy water from rivers.

At some locations, sand is a nuisance — for example, where it blocks navigation channels and where dunes encroach on existing developments. Under these circumstances, sand mining achieves two ends: the removal of the nuisance and an increase in the availability of sand.

Mineral extraction: hydrocarbons

The extension of sovereignty up to 200 nautical miles offshore has stimulated the exploitation of hydrocarbon resources such as oil and gas. Countries offer, for lease or sale, exploration rights in blocks in the zone extending to the edge of the continental shelf. Today, locations such as the Gulf of Mexico, the Persian Gulf, the Bight of Biafra in west Africa, and the North Sea are major contributors to oil and gas production. Such is the demand for hydrocarbons that few areas that have the geological potential to hold reserves are left unexplored. As technology advances, drilling in deeper and rougher water becomes practical.

Both offshore and onshore facilities are developed to service the energy industry. Terminals, where hydrocarbons are transferred, storage tanks, pipelines and refineries are built. Supertankers (VLCCs Very Large Crude Carriers) have particular demands — for example, a water depth of 30 m or more and not less than a 2 km stretch of open water in which to turn.

Hydrocarbon exploitation has both positive and negative impacts. Positive impacts include:
- employment creation
- wealth creation
- manufacture of products (e.g. plastics enabling a higher standard of living)

Negative impacts include:
- oil spills
- ecosystem disturbance
- visual impact

The socioeconomic impact on local communities can be either positive or negative, depending on the circumstances.

Oil spills

The release of oil into the environment comes about in a number of ways. Dramatic spills result from damage to tankers. However, this accounts for only 5% of leaked oil. Most comes from spills during routine operations, such as when oil is transferred between tanker and terminal.

What happens to oil when it spills into salt water is a complex process. It depends partly on the type of oil and partly on the sea and atmospheric conditions. In general, slick development goes through three phases:
- initial gravity dispersion, i.e. sinking of some oil (0–5 minutes)
- viscous advection, i.e. horizontal spreading of treacle-like oil (up to 40 hours)
- surface-tension spreading, i.e. dispersion of some oil across the surface (up to 150 days)

Processes of evaporation, solution, emulsification, oxidation and aeration take place. When the oil is reduced to low concentrations, it is biodegraded by microbes.

Direct, short-term effects can be visually and ecologically dramatic. An impact on one trophic level has significant effects throughout an ecosystem. Mass mortalities of some species result in a loss of biodiversity. Levels of resistance to oil vary among species, depending on such characteristics as feeding behaviour. Diving birds, for example cormorants and guillemots, can be affected severely.

Oil spills on coral reefs

Corals are sedentary organisms, so oil spills are particularly threatening. Some of the most active regions in terms of oil extraction, processing and transport coincide with extensive reefs. The Arabian Gulf, the Red Sea and the Panama Canal are locations where oil and coral come into contact. The accidental release of 50000 tonnes of crude oil from Panama's Isla Payardi refinery in 1986 led to the contamination of coral reefs, seagrass beds and mangroves. Immediate effects were obvious, with high mortality for many species, for example bivalves living among the mangrove roots and the coral itself. Oil accumulated and became incorporated into sediment held within the mangrove ecosystem. This was released gradually in the years following the spill, contaminating local ecosystems. Sediment became subject to increased wave action because there was less protection from the damaged reefs and seagrass beds. Ongoing studies indicate that it can take over 25 years for an ecosystem to recover from the impact of an oil spill.

Preventing and dealing with oil spills

Methods of coping with oil spills have had to be improved because of the significant increase in the size of tankers since the 1960s. In 1967, the *Torrey Canyon* ran aground on the Seven Stones reef, off Land's End, Cornwall, spilling 120000 tonnes of crude oil. In 1978, the *Amoco Cadiz* was wrecked off the Brittany coast with the release of 230000 tonnes of crude oil.

Prevention techniques involve improvements in ship design and in methods of handling oil. Preparing for spills involves the assessment of risk and of the likely impact on the coastal zone, particularly in locations near to oil installations and along busy transport routes. The persistence of oil on different types of shore (e.g. rock, sand, marsh) and the sensitivity of ecosystems to oil pollution are assessed and mapped. The reaction to oil spills depends on the resources of the relevant authorities. The level of funding is important, as is the provision of machinery and manpower. A number of techniques exist for clearing up after an oil spill. On water, floating booms are used to contain surface oil that can then be pumped into tankers. Chemical dispersants can be effective, but may be more toxic than the oil itself.

Once the oil washes up on the shore, its physical removal and the disposal of bird and other animal carcasses becomes the priority. Surface oil can be collected by scraping. After the *Amoco Cadiz* incident, 35 000 French military personnel were involved for a month collecting surface deposits. However, once oil seeps into coastal sediments, cleaning becomes difficult. Wholesale removal is often impractical and can have a destabilising effect on the equilibrium of sediment budgets. Techniques such as the introduction of bacteria that consume oil, and/or irrigating or aerating the sediment to accelerate the breakdown of the oil, can be attempted.

Tourism and recreation

Over the past 200 years, tourism and recreation have developed into major economic activities on the coast. As disposable income, personal mobility and paid holiday increase, growing numbers of people visit the seaside to enjoy its resources. By the end of the nineteenth century, in most MEDCs, seaside holidays became part of the annual rhythm of peoples' lives. Sea-bathing moved from being a small-scale, fashionable and therapeutic pursuit of the rich, to mass tourism. All sectors of society now participate, with some people being able to exploit coastal resources for tourism on any continent — even Antarctica.

Coastal resources for tourism and recreation

As with any economic activity, tourism and recreation require resources in order to operate. The physical 'raw materials' are:
- relief (e.g. steep cliffs, low-lying land, slope of offshore gradient)
- beach material (e.g. pebbles, sand)
- water (e.g. nearshore water flows (rip currents/gentle currents), temperature, quality)
- climate (e.g. Mediterranean, arctic)
- ecosystem (e.g. estuary, coral reef)

Different combinations of physical resources often result in the development of different types of tourism. Where easy access to the shore, a sand beach and a climate with a hot and dry season are found together, beach- and water-based activities flourish (Figure 6.14). Well-known examples include the resorts along the southern Californian, eastern Australian and western Mediterranean coastlines.

High biodiversity in some coastal locations offers a number of recreational opportunities. Estuaries, for example the Exe, attract ornithologists; the clear water and spectacular reef growths in the Red Sea attract divers from around the world.

As well as physical resources, tourism also seeks to use human resources — for example:

- cultural attractions (e.g. theatres, nightclubs, restaurants)
- heritage resources (e.g. piers, architecture, preserved railways)

Some physical resources acquire cultural or heritage status. For example, large stretches of the coast of England and Wales are owned by the National Trust or are designated as national parks.

| Figure 6.14 | **Beach development, Vina del Mar, Chile** |

Michael Raw

Activity 4

(a) Using either a 1:50000 or 1:25000 OS map, sketch a suitable length (about 10–15 km) of coastline.

(b) Using appropriately designed symbols, plot the location of (i) the physical resources and (ii) the human resources that tourism and recreation might exploit.

(c) Annotate your map with comments explaining the locations of the resources that could be exploited for tourism and recreation.

Conflict among recreational activities

Within the coastal zone, the diversity of recreational activities, and other activities not associated with this sector, often leads to conflicts. The same stretch of beach or water may be contested by incompatible activities. For example, power-boating, jet-skiing and water-skiing use water space in ways that do not mix with swimming, sailing and fishing. Noise and high-energy wakes from mechanically powered craft can disturb wildlife and erode shorelines.

In many coastal locations, the growth in demand from tourism and recreation causes significant problems. After assessment of the issues, zoning of both beaches and water space may be implemented. This approach tries to allocate space to activities either exclusively or with a degree of limitation. The less compatible activities are, the more separation they require.

Zoning in the Great Barrier Reef Marine Park, Australia

Stretching more than 2300 km along Australia's northeastern coast is the Great Barrier Reef. This is the world's largest coral reef and includes over 1000 islands. Marine life flourishes along the reef. Such is the area's importance that it is a **World Heritage Area**. A variety of human activities take place along the reef, including:
- fishing, both commercial and recreational
- boating, including sailing and power boats
- mineral extraction
- snorkelling and diving
- beach tourism
- scientific research

Some of these uses are compatible with each other; others are not. Zoning is seen as offering viable management of the reef. It attempts to balance the interests and needs of the ecosystem with human activities.

The zone categories have varying degrees of restriction and exist throughout the entire reef. They are shown in Table 6.3.

Table 6.3 Zoning in the Great Barrier Reef Marine Park, Australia

Zone	Percentage of park area	Degree of restriction
General use	32.5	Most activities allowed
Habitat protection	28.0	
Conservation	1.5	
Buffer	3.0	
Scientific research	1.0	
Marine National Park	33.0	
Preservation zone	1.0	Nothing allowed

On a more local scale, zoning of recreational activities takes place in some parts of the Park where tourist pressure is particularly intense. Green Island lies off the coast of mainland Australia, near the settlement of Cairns. This is a low, tree-covered coral island of about 12 ha. Its maximum height above sea level is 4.5 m and it can be covered by storm surge. The first jetty was built in 1906. The island's current popularity stems from its natural environment and its convenient location, having relatively easy access from Cairns. These features led to, and are now reinforced by, commercial tourist facilities. Mostly, day visitors come to swim, snorkel, dive or view the reef ecosystem from glass-bottomed boats. Overcrowding and the incompatibility of some activities have led to a zoning scheme (Figure 6.15). The zoning attempts to protect the natural assets of the area — the very reason why people wish to visit.

Figure 6.15 **Zoning plan, Green Island, Great Barrier Reef Marine Park**

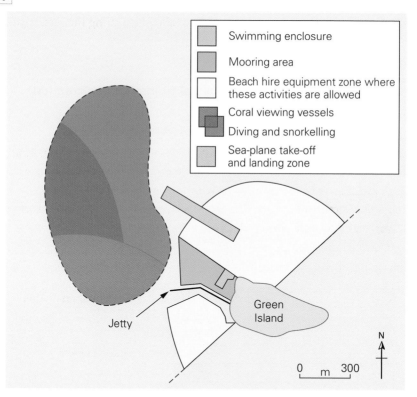

Resorts on land

On land, resorts take up space. The issues arising here are similar to the growth of settlements anywhere, for example the use of greenfield sites for housing and the degradation of buildings when a resort declines in popularity.

The siting of campsites and, in particular, caravan sites, can also cause considerable conflict. Their visual impact can be locally significant (Figure 6.16). However, with careful screening some, if not all, of their impact can be lessened.

Figure 6.16 Caravan site, Devon

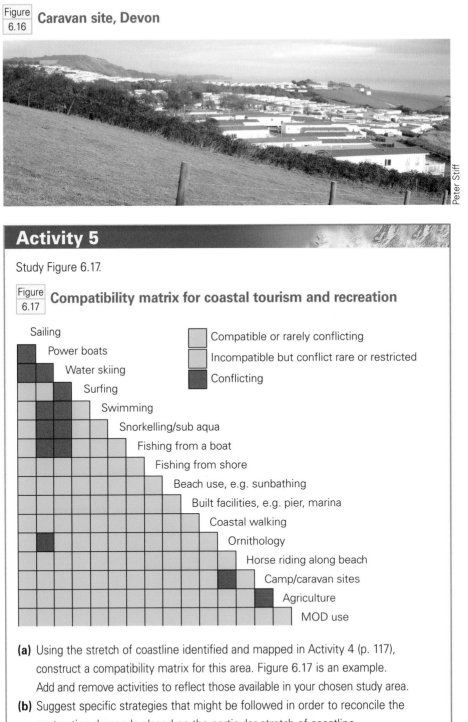

Peter Stiff

Activity 5

Study Figure 6.17.

Figure 6.17 **Compatibility matrix for coastal tourism and recreation**

Sailing
  Power boats
    Water skiing
      Surfing
        Swimming
          Snorkelling/sub aqua
            Fishing from a boat
              Fishing from shore
                Beach use, e.g. sunbathing
                  Built facilities, e.g. pier, marina
                    Coastal walking
                      Ornithology
                        Horse riding along beach
                          Camp/caravan sites
                            Agriculture
                              MOD use

☐ Compatible or rarely conflicting
☐ Incompatible but conflict rare or restricted
■ Conflicting

(a) Using the stretch of coastline identified and mapped in Activity 4 (p. 117), construct a compatibility matrix for this area. Figure 6.17 is an example. Add and remove activities to reflect those available in your chosen study area.

(b) Suggest specific strategies that might be followed in order to reconcile the contrasting demands placed on the particular stretch of coastline.

7 Managing the coastal zone

The relationship between humans and the coastal zone has often been uneasy. The coast is attractive to people and its processes and landforms have been deliberately modified in order to:

- restore ecological and geomorphologic equilibrium to coastal locations that have been degraded by human activity, for example sand dunes and beaches
- reduce or eliminate hazards that occur in the coastal zone, for example flooding by seawater and cliff failure
- develop the resources of the coastal zone for human activity

A range of management techniques has been applied to coastlines, with varying degrees of success. However, what is meant by 'success' is being redefined because some techniques that were once viewed positively have been found to have effects that impact negatively on the coastal system.

Hard-engineering approaches

Over most of the past 150 years, the construction of timber, concrete or rock structures, often on a large scale, was thought to be the best way to combat the effects of coastal erosion and flooding. It was assumed that these structures would be able to resist wave action.

Preventing marine erosion and sub-aerial weathering

Sea walls are designed to prevent shoreline erosion by reflecting wave energy. Initially, these were vertical structures but their design has changed gradually. Curves and steps, which dissipate wave energy to some degree, are now incorporated into their design (Figure 7.1). When they were first constructed, one of the unforeseen consequences was the turbulence set up at their bases. This scours away sediment, leading to undermining and eventual collapse of the wall.

Figure 7.1 Sea wall design

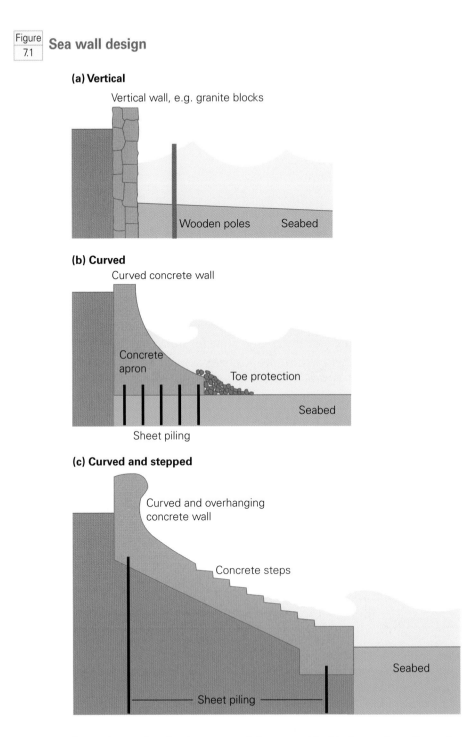

(a) Vertical

Vertical wall, e.g. granite blocks

Wooden poles   Seabed

(b) Curved

Curved concrete wall

Concrete apron

Toe protection

Seabed

Sheet piling

(c) Curved and stepped

Curved and overhanging concrete wall

Concrete steps

Seabed

Sheet piling

Designs have changed as further research has provided information about the movement of water and sediment in front of a wall (see Figures 7.2 and 7.3).

Figure 7.2 Curved sea wall, Slapton Ley, Devon

Michael Morrish

Figure 7.3 Sloping sea wall, Dawlish Warren, Devon

Peter Stiff

Activity 1

(a) On a copy of Figure 7.1, annotate the sea-wall cross sections to describe and explain the advantages and disadvantages of each type of wall in resisting wave attack.

(b) On the same diagram, describe and explain the problems that might arise from processes on the landward side of the wall. Suggest possible solutions to these problems.

Particular issues arise where a sea wall ends. Waves refract round the wall end leading to erosion of the shore. This process is called **flanking** (Figure 7.4).

| Figure 7.4 | **Erosion caused by 'flanking' at the end of a sea wall** |

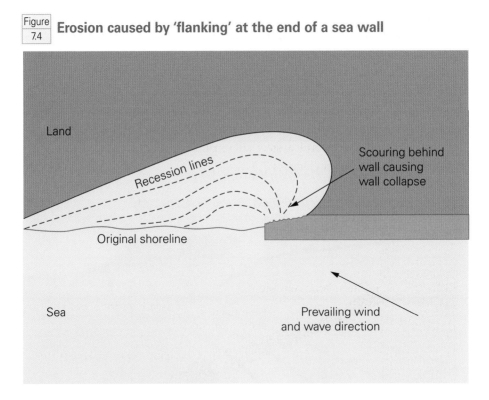

The Exmouth town wall

Exmouth is one of many resorts in the east Devon coastal zone (see Figure 7.17, p. 140). The first sea wall, built in the mid-nineteenth century, consisted of masonry supported on timber piles sunk into the sand. It protected the developing seaside resort's promenade, gardens, hotels and boarding houses.

By the beginning of the twenty-first century, various issues concerning the wall had arisen:

- Natural eastwards migration of the River Exe channel has been prevented by the sea wall.
- The beach at Exmouth has steepened as the high-energy river has flowed closer to the shore.
- The beach height now varies by as much as 1–2 m over one or two tides.
- The beach level has dropped below the top of the wooden piles.
- The wooden piles and masonry foundations are now exposed to erosion and undermining from strong currents, waves and sub-aerial processes.

The nineteenth-century sea wall itself is in good condition, but the exposure of its foundations threatens its stability. The wall is essential to protect the sea front and the town centre from flooding; it is increasingly necessary because of the rising sea level. In 2004, strong southeasterly storms and high spring tides resulted in partial subsidence of the wall. As a temporary measure, rock armour was placed in front of the wall.

Currently, two phases of management are in place:

- Immediate measures involve partial removal of the existing sea-wall foundation and its replacement with reinforced concrete faced with stone, with a stepped apron and a steel-sheet pile toe extending well below the typical low-beach level (Figures 7.5 and 7.6).

Figure 7.5 **Cross section of refurbished sea wall, Exmouth**

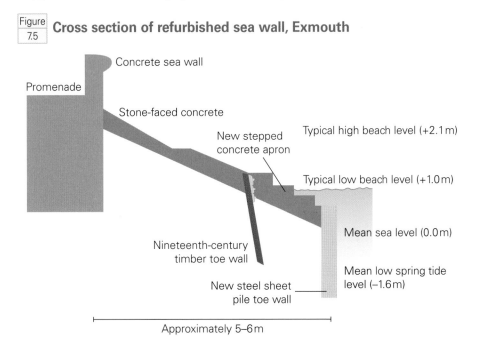

Promenade
Concrete sea wall
Stone-faced concrete
New stepped concrete apron
Typical high beach level (+2.1 m)
Typical low beach level (+1.0 m)
Mean sea level (0.0 m)
Nineteenth-century timber toe wall
New steel sheet pile toe wall
Mean low spring tide level (–1.6 m)
Approximately 5–6 m

Figure 7.6 **Reconstruction of Exmouth's sea wall**

Peter Stiff

- Longer-term measures to restore and maintain the beach at a higher level involve the Exe Estuary Coast Management Study in determining the most appropriate measures, taking into account the possible impact on Dawlish Warren and the SSSI at Exminster Marshes.

Other hard-engineering methods

Protection of the shore increasingly uses 'hard' techniques that absorb and dissipate wave energy.

Gabions are wire-framed cubes filled with rock pieces. They can be stacked vertically and horizontally (Figure 7.7). The air spaces between individual rock pieces are accessible to the sea, so the energy of an advancing wave is absorbed within the gabion.

Figure 7.7 **Gabion wall, Norfolk**

Michael Raw

Revetments are sloping, open structures made of various materials, with wood and concrete being commonly used. Advancing waves break through the structure and the wave energy is dissipated. Revetments reflect less energy than solid walls and generally cause less scouring of the beach. Sediment accumulates behind the revetments, providing further protection. However, in order to offer the degree of protection equivalent to a wall, revetments have to cover a larger area.

Some 'hard' structures involve pre-cast concrete shapes. Two of the most common are **dolos** (orthogonal T-shapes) and **tetrapods** (four-legged shapes). Piles of these shapes are used to protect shore areas, often where industrial plant, for example an oil or gas installation, is located close to the sea. In addition to these man-made types of **rip-rap,** large boulders are placed seawards of locations that require protection. Granite or basalt rocks are often used, because of their resistance to both marine and sub-aerial attack. Natural rip-rap often involves material that is geologically different from that available locally (Figure 7.8).

Rip-rap and sea wall, Morecambe, Lancashire

Michael Raw

Cliff stabilisation involves a number of techniques that are designed to prevent slope failure. Cliff-foot strategies are used to try to minimise wave impact; other techniques are aimed at reducing or preventing the effects of sub-aerial processes, particularly where unconsolidated material, for example clay or shale, is present. Some cliffs are regraded by material being removed from the upper part of the slope and/or material being added to the lower part of the slope. This reduces the overall slope angle and so decreases the chance of mass movement, such as rotational slip.

Activity 2

Some cliff-stabilisation techniques are appropriate to more than one rock type. State and explain the type of geology and accompanying cliff profile likely to benefit from each of the following techniques:
- netting
- pinning with metal bolts
- encouraging the growth of vegetation
- draining water from the cliff
- lowering the angle of the cliff slope

Hard-engineering techniques have a combination of the following disadvantages:
- high cost
- adverse visual impact
- reduced access to, and use of, the beach

Embankments or **dykes** were one of the first management techniques employed against the sea. There is evidence of their use in Roman times. They are made of unconsolidated material, such as clay, and are built above the mean spring high tide level. Their primary aim, therefore, is to prevent the sea from flooding low-lying areas. They are found lining estuaries and on the landward side of salt marshes. If sea level rises, **coastal squeeze** can occur between the rising high tide level and the embankment, reducing the width of the salt marsh. Therefore, there is a loss of habitat and a reduction in the protection that an area of marsh can give to the shore on its landward side (Figure 7.9).

Figure
7.9
Coastal squeeze

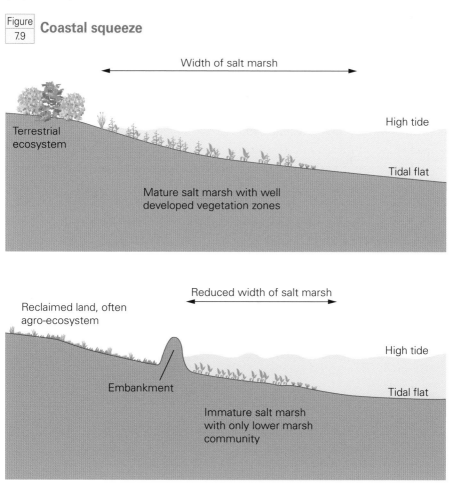

Promoting sediment build-up

As waves approach the shore, **detached breakwaters** and **artificial reefs** intercept and dissipate wave energy. Stone walls, concrete (solid wall or pre-cast shapes), boulders and even used tyres offer protection to the shore immediately behind them. Patterns of wave refraction/diffraction and the lower energy conditions created in the lee of the structure lead to sediment accumulation. Engineers hope that a beach plan with sustainable equilibrium will develop. At some locations, a tombolo grows out from the shore to the breakwater. The important influences on breakwater design are:

- the length of the gap between the breakwater ends
- the length of the breakwaters
- the distance of the breakwaters from the shore

It is these dimensions that engineers model in order to avoid undesirable effects such as waves regaining energy after passing the breakwater. This happens if the structure is placed too far offshore. Cost and aesthetic considerations also play a part in design and location.

Attached breakwaters protect harbours, docks and moored vessels from wave energy.

Groynes were one of the earliest management techniques designed to encourage sediment accumulation (Figure 7.10). There are records of their use as early as the seventeenth century.

| Figure 7.10 | **Groyne and sediment accumulation, Cuckmere Haven, Sussex** |

Direction of longshore drift

Build up of sediment against a groyne

Michael Morrish

Groynes comprise a wall constructed of wood, concrete or boulders (rip-rap) extending perpendicular from the beach. Several groynes are usually placed next to each other in a groyne field. A critical aspect of their use is the ratio between groyne length and spacing (Figure 7.11). On sandy beaches, a ratio of 1:4 seems to be most effective; on gravel and shingle beaches, 1:2 is considered optimal.

Figure 7.11 **The effects of groyne spacing**

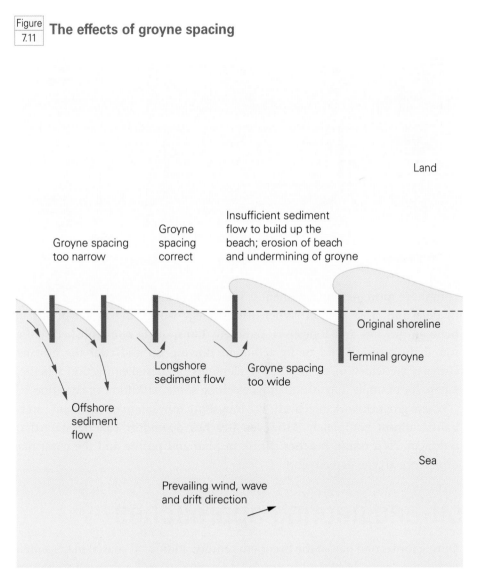

In some locations, angled groynes are used to counteract offshore sediment transport by rip currents (Figure 7.12).

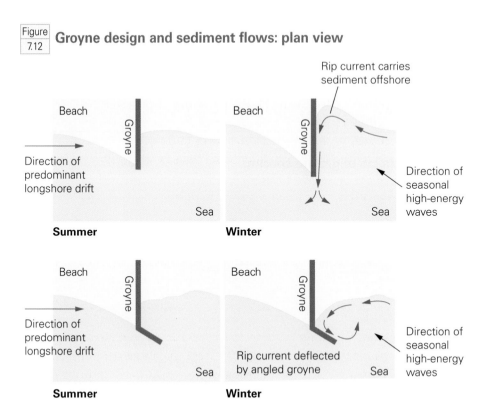

Figure 7.12 **Groyne design and sediment flows: plan view**

Longshore movement of sediment is interrupted, and a beach builds up both in plan and cross section. When this strategy works, the beach builds up between groynes and longshore sediment transport is re-established, either round the seaward end or by sediment moving up and over the groynes. Groynes are widely used as effective elements of coastal protection schemes. Japan, for example, has 10000 groynes along some 32000km of coastline.

Where groynes work effectively, the longshore movement of sediment can be halted almost completely. However, beaches downdrift can be starved of sediment. As a result, beaches shrink in plan and profile and the coast may experience accelerated erosion.

Soft-engineering approaches

During the second half of the twentieth century, additional coastal management techniques were developed to try to use natural coastal processes to prevent shoreline erosion. The focus was on building up beaches, so that wave energy would be absorbed by greater accumulations of sediment. The theory is that not only is land protected, but the coastline has a more natural appearance.

Beach nourishment

Beach nourishment requires sediment to be brought into the area to build up the plan and profile of the beach. Key elements in beach nourishment schemes are:

- a source of sediment nearby
- the impact of sediment removal by dredging/mining
- calibre of sediment equal to, or a little coarser than, existing sediment (it should match the natural beach sediment)
- the techniques used to retain new sediment *in situ*

> ## Activity 3
>
> For each of the key elements mentioned above, suggest *how* and *why* it has to be taken into account when beach nourishment is carried out.

A significant issue regarding beach nourishment is the way that sediment is redistributed once it has been deposited. Wave energy can soon rework sediment placed on a beach, leading to the perception that the time, effort and money involved have not been worthwhile. However, beach systems tend to adjust automatically through dynamic equilibrium. The new sediment added to the beach is soon sorted by size. Sometimes, it is moved just offshore. Material may not be lost from the offshore zone, but stay close by, from where it becomes part of the sediment system and plays a role in building up a beach. At some locations, research suggests that placing sediment in shallow water close to the shore and allowing the waves and currents to redistribute it, is an appropriate strategy. The nature of beaches is dynamic, so nourishment schemes are planned for a sequence of replenishment lasting years.

Many beach systems benefit from nourishment, including those at Copacabana in Rio de Janeiro and Miami in Florida. The latter is designed to withstand storm surges associated with hurricanes, in order to protect valuable resort and residential developments.

Artificial dune construction

Artificial dune construction is found along sandy coastlines where no natural dune systems exist. It is used on lowland coasts, for example in the Netherlands and in parts of the east coast of the USA, as a defensive measure. Structures designed to slow down wind speed are erected along the coast. Materials such as brushwood, wooden palings (thin lengths of wood wired together with gaps between them) or woven fabric (often plastic based) are used. In general, the aim is to reduce airflow by about 50% and to encourage sand accumulation on either side of the fence. It is important that the fence is permeable to allow filtration of the moving air with its load of saltating sand. A solid fence would set up too much turbulence and reduce sand deposition. Once the fence is almost covered, either it can be lifted to a new height or another fence can be installed to continue sand accumulation. Where there is a steady supply of sand and sufficient energy from wind, rates of accumulation of about $1\,\mathrm{m\,yr^{-1}}$ can be achieved. This method starts dune building artificially. For the scheme to be sustainable, vegetation has to be planted to stabilise the system (Figure 7.13).

Figure 7.13 **Artificial dune construction on barrier islands, North Carolina**

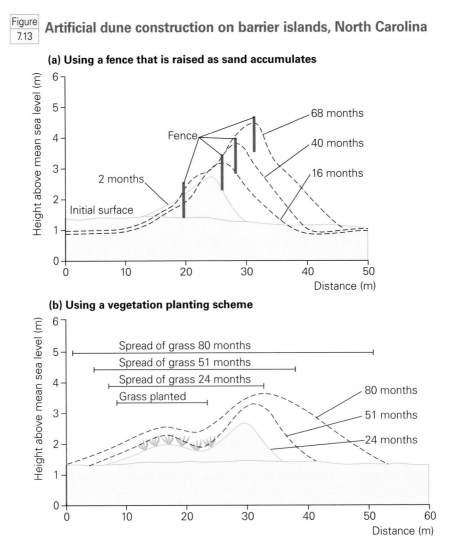

(a) Using a fence that is raised as sand accumulates

(b) Using a vegetation planting scheme

Dune stabilisation

Dune stabilisation is carried out where natural dune systems are threatened. Repairing gaps in ridges through a combination of fences and vegetation planting is preferable. However, the invasive technique of moving sand by mechanical means, for example using earthmoving equipment, is often tried. At Braunton Burrows, north Devon, management of the dune system now includes the deliberate creation of blowouts to release sand for aeolian transport. It is hoped that the mobile sand will accumulate around marram plants. These need fresh sand to build up around their stems; without it, they will die.

Managed retreat

Coastal management schemes have to address problems caused by rising sea levels. The rate of sea level rise is accelerating, which adds further pressure to these schemes. Given the high, and growing, cost of schemes to prevent erosion and flooding, alternative approaches are being tried. One of these techniques is called **managed retreat** or **coastal realignment**. This means allowing the sea to come into areas that were protected previously. In time, salt marshes and mudflats develop to form a zone that provides a natural defence to the 'new' shore landwards of the abandoned area. It is hoped that such schemes will be more sustainable than some of those based on hard-engineering techniques. Anticipated consequences of such an approach include:

- minimisation of the risks to the human population
- relocation of coastal settlement and economic activity further inland
- a reduction in the costs of protection
- the encouragement of 'natural' protection, for example mudflats and salt marsh, to counter coastal squeeze
- the production of habitats that have high ecological value, for example mudflats and salt marsh

Such schemes are relatively new. They are controversial and their full impact is not yet known.

Managed retreat in Essex: Abbotts Hall Farm

Southeast England is experiencing both eustatic rise and isostatic sinking. These processes are placing considerable pressure on stretches of low-lying coast. Over several centuries, many tens of kilometres of embankments have been built as protection. In southeast England, sea level is currently rising at about $6\,\mathrm{mm\,yr^{-1}}$.

In Essex, some 40% of all salt marsh has been lost to coastal squeeze in the last 25 years. At Abbotts Hall Farm on the Blackwater estuary, a managed retreat scheme was implemented in 2002. Five breaches were made in sea walls, allowing salt water to cover former arable fields (Figure 7.14). It was not just a case of making breaches; flanking walls were required at the edges of the scheme to prevent salt water spreading into farmland that was to be retained. Some channels were excavated to encourage the movement of salt water into the area. The hope is that the flooded areas will develop into mature salt marsh that will act to absorb wave and current energies.

Lessons have been learnt from this project. They include:

- the best shape and size for the man-made channels
- the measures needed to reduce the transport of silt from the flooded fields back into the estuary by ebb tides
- how to manage the land before flooding so as not to release agrochemicals.

Figure 7.14 **Abbotts Hall Farm, Essex**

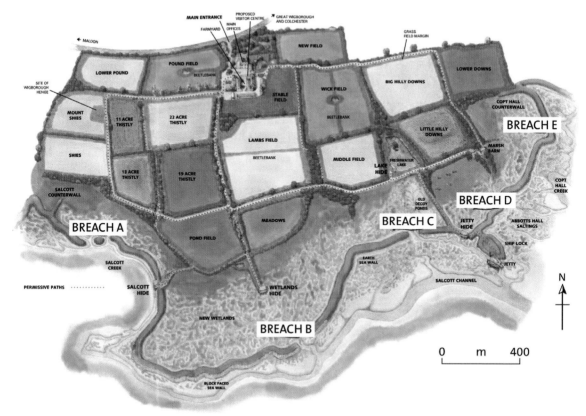

Activity 4

(a) Using a variety of sources, such as OS maps, newspaper articles and websites, describe and explain the type of coastline best suited to a managed retreat approach.

(b) Using the same set of sources, assess the extent to which different stakeholder groups are likely to support or oppose managed retreat as an option.

New frameworks for coastal management

In previous chapters, we have seen that the coastal zone operates as a system and that within this overall structure there are smaller sub-systems. This is true both in terms of the processes that are active along the coast and the geographic scale of investigation.

In the past, coastal management tended to deal with a single issue at a time. At one location, sediment loss from beaches was the concern; at another, it was an eroding cliff; elsewhere, it was the reclamation of tidal flats. Each issue tended to be dealt with by an individual authority. Although the original issue was often resolved, unintended consequences frequently arose. For example, cliff protection along one stretch of coast often led to sediment starvation and increased erosion elsewhere.

Integrated coastal management overcomes this piecemeal approach and is now the preferred strategy. It is recognised that the geographical context of a coastline includes not only the cliffs and beaches, dunes and marshes, estuaries and nearshore areas, but also the river catchments draining into the coast. The importance of bringing together the groups and individuals who have a stake in the coastal zone is appreciated.

Sediment cells are accepted as being important parts of the coastal system and they are used as the basis for management (Figure 7.15). Such an integrated approach is vital. When planning how to respond to the accelerating rise in sea level, it has the strong support of the **Intergovernmental Panel on Climate Change (IPCC)**.

Figure 7.15 **Major sediment cells for England and Wales**

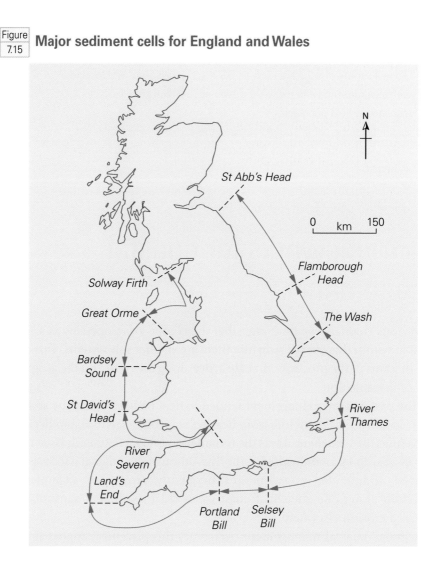

Increasingly **sustainable** management of the coast is the primary aim. This means using coastal areas for the benefit of the present population while maintaining the potential of coastal systems to meet the aims and aspirations of future generations. However, this admirable aim may be ill-defined and is not easy to attain.

A key advantage of a sustainable approach is that it alters perceptions and ways of thinking about issues. The consequence is that management approaches now tend to take a longer-term view and are more **holistic**. Clearly, it is impossible to resolve all coastal issues. Plans should recognise that the environment cannot be managed in the rigid and unsustainable ways we have seen up to now. Today, it is understood that coastal management needs to be flexible and should include human, as well as physical, elements.

Shoreline Management Plans (SMPs)

Towards the end of the twentieth century, integrated plans for the management of the coastline of England and Wales were given an official structure, based largely on the sediment-cell systems (Figure 7.16). Responsibility for each area rests with a combination of organisations, for example the Environment Agency, the Department for Food and Rural Affairs (DEFRA), local maritime authorities, local councils, English Nature and the National Trust.

| Figure 7.16 | **Coastal management groups and major water and sewerage companies in England and Wales** |

Once information has been gathered, consultation has taken place and a draft policy has been proposed, a plan for the length of shoreline is proposed. This includes an overall strategy. In addition, particular policies for individual or discrete sections of coast — **Management Units** — are included. These can range from several kilometres of a dune system to less than 1 km for an individual village. Four principal policy options are considered for each unit:

- do nothing
- hold the existing coast line
- advance the existing coastline seawards
- managed retreat from the existing coastline

SMPs are not intended to put in place a strategy that is fixed forever. They can be modified as the coastal system changes. Some of the original plans are already being adjusted as more detailed knowledge about sea-level rise is gathered and changing weather patterns in the form of increased storm energy are experienced.

Shoreline Management Plans: East Devon

This stretch of coast is part of the Lyme Bay and South Devon coastline group and consists of three Management Units (Figure 7.17).

| Figure 7.17 | **Part of the east Devon coastal zone** |

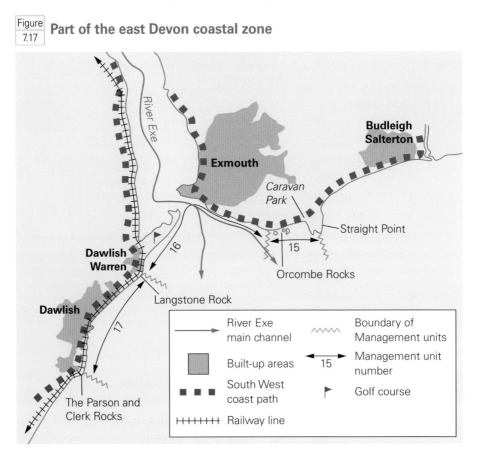

The Exe estuary dominates the coastal zone, which complicates the wave, tide and sediment-flow patterns. Land use is predominantly residential, with some agricultural and recreational uses. Sea use includes fishing, some commercial and tourist-related activities and recreation — for example, the marina at Exmouth. The eastern Unit marks the beginning of the Jurassic Coast, a World Heritage site because of its internationally important geology and geomorphology. The Exe estuary is an internationally important habitat for bird life, in both summer and winter.

The three Units have been given the following strategic options and specific objectives:

- Management Unit 15 stretches from Straight Point to Orcombe Rocks. The strategic option is 'do nothing'. The specific objectives are:
  - to protect the caravan park
  - to maintain the continuity of the South West coast path
  - to maintain the bathing–beach quality
  - to maintain the recreational and amenity values of the area
  - to maintain the integrity of the nationally and internationally designated sites
- Management Unit 16 stetches from Orcombe Rocks to Langstone Rock. The strategic option is to 'selectively hold the line'. The specific objectives are:
  - to protect people and property in Exmouth
  - to protect the railway line
  - to not restrict navigational access to the estuary
  - to maintain bathing–beach quality
  - to protect the golf course
  - to protect at-risk listed buildings and conservation areas
  - to ensure that the shellfish industry is not affected adversely
  - to maintain the continuity of the South West coast path
  - to maintain the integrity of the nationally and internationally designated sites
  - to maintain the natural coast protection and flood defence elements in the system
- Management Unit 17 stretches from Langstone Rock to The Parson and Clerk Rocks. The strategic option is to 'selectively hold the line'. The specific objectives are:
  - to protect people and property in Dawlish
  - to protect the railway line
  - to maintain the continuity of the South West coast path
  - to maintain the integrity of the nationally and internationally designated sites
  - to protect at-risk listed buildings and conservation areas
  - to not adversely affect the recreational and amenity elements of the area

The options and objectives of the SMP are to be reviewed at 5-yearly intervals.

Activity 5

(a) Research specific Shoreline Management Plans and individual Management Units for a particular stretch of coastline.

(b) Evaluate the range of Specific Objectives for each Management Unit in terms of their impact on the different stakeholders.